Advanced Series in Agricultural Sciences 17

Advanced Series in Agricultural Sciences

Volumes already published in the series

Volume 1: *A. P. A. Vink*
Land Use in Advancing Agriculture

Volume 2: *H. Wheeler*
Plant Pathogenesis

Volume 3: *R. A. Robinson*
Plant Pathosystems

Volume 4: *H. C. Coppel, J. W. Mertins*
Biological Insect Pest Suppression

Volume 5: *J. J. Hanan, W. D. Holley, K. L. Goldsberry*
Greenhouse Management

Volume 6: *J. E. Vanderplank*
Genetic and Molecular Basis of Plant Pathogenesis

Volume 7: *J. K. Matsushima*
Feeding Beef Cattle

Volume 8: *R. J. Hanks, G. L. Ashcroft*
Applied Soil Physics

Volume 9: *J. Palti*
Cultural Practices and Infectious Crop Diseases

Volume 10: *E. Bresler, B. L. McNeal, D. L. Carter*
Saline and Sodic Soils

Volume 11: *J. R. Parks*
A Theory of Feeding and Growth of Animals

Volume 12: *J. Hagin, B. Tucker*
Fertilization of Dryland and Irrigated Soils

Volume 13: *A. J. Koolen, H. Kuipers*
Agricultural Soil Mechanics

Volume 14: *G. Stanhill*
Energy and Agriculture

Volume 15: *E. A. Curl, B. Truelove*
The Rhizosphere

Volume 16: *D. P. Doolittle*
Population Genetics: Basic Principles

Volume 17: *A. Feigin, I. Ravina, J. Shalhevet*
Irrigation with Treated Sewage Effluent

Volume 18: *D. Gianola, K. Hammond*
Advances in Statistical Methods for Genetic Improvement of Livestock

A. Feigin I. Ravina J. Shalhevet

Irrigation with Treated Sewage Effluent

Management for Environmental Protection

With 46 Figures and 66 Tables

Springer-Verlag
Berlin Heidelberg New York
London Paris Tokyo
Hong Kong Barcelona

Dr. AMOS FEIGIN
Institute of Soils and Water
ARO, The Volcani Center
P.O. Box 6
Bet-Dagan 50-250, Israel

Prof. Dr. ISRAELA RAVINA
Faculty of Agricultural Engineering
The Technion
Haifa, Israel

Dr. JOSEPH SHALHEVET
Institute of Soils and Water
ARO, The Volcani Center
P.O. Box 6
Bet-Dagan 50-250, Israel

ISBN-13: 978-3-642-74482-2 e-ISBN-13: 978-3-642-74480-8

DOI: 10.1007/978-3-642-74480-8

Library of Congress Cataloging-in-Publication Data. Feigin, A. (Amos), 1930– . Irrigation with treated sewage effluent : management for environmental protection / A. Feigin, I. Ravina, J. Shalhevet. p. cm. ––(Advanced series in agricultural sciences ; 17) Includes index. 1. Sewage irrigation. I. Ravina, I. (Israela) II. Shalhevet, Joseph. III. Title. TD760.F45 1991 628.3'623––dc20 90-10199

Softcover reprint of the hardcover 1st edition 1991

31/3145(3011)-543210 – Printed on acid-free paper

Preface

The growing volume of municipal wastewater resulting from the rapidly increasing world's urban population creates severe environmental and health hazards, such as uncontrolled flows of sewage-polluting surface and groundwaters, soil, and agricultural products. Different technologies have been developed to prevent this type of environmental degradation, among which the use of land as a living filter for removing pollutants from wastewater is especially popular. This technology, which is based on the soil-vegetation complex, is a viable, environmentally sound and costeffective method for the disposal of wastewater.

Irrigation is one of the main land treatment approaches, and as other land treatment methods (e.g. infiltration-percolation), it is considered an efficient method of waste disposal. Numerous publications provide information on different components of existing sewage disposal systems. However, wastewater should also be seen as an alternative source of irrigation water, especially in dry regions, where reclaimed wastewater is sometimes the only source of irrigation water for agricultural development. The increasing use of sewage effluent for irrigation is documented in many relevant publications, many of which are reviewed and discussed in this book.

The present volume describes the main components of effluent-soil-plant systems necessary for the development of appropriate irrigation-fertilization-cropping management systems and cites representative case histories recorded in Israel and elsewhere. Comprehensive information has been provided on the following topics: (1) sources, treatment methods and properties of treated sewage effluent; (2) the main processes of different effluent constituents (e.g. salt, nutrients, trace elements, organisms) in soil-plant systems; (3) irrigation and fertilization management; and (4) irrigation systems for effluent irrigation.

The objective of the book is to provide the reader with a comprehensive overview of the principles and practices of using sewage effluent for irrigation. Although all aspects of irrigation management are discussed, special attention is given to the practices required to overcome the specific limitations imposed by the quality of the water, i.e., pathogenic microorganisms, high salinity and sodicity, high nutrient content, and under specific circumstances, the presence of toxic levels of microelements, organic pesticides, and suspended solids.

The book is not intended to be a comprehensive scientific review of published literature related to effluent irrigation, although major contributions in this field have been cited. Naturally, we have relied heavily on the Israeli experience to provide examples and demonstrations of some of the ideas and principles. The book is intended especially for the field irrigation manager (agronomists, engineers and farmers), but it can also be used as a general reference for students and other specialists interested in the use of sewage effluent for irrigation. The reader will find valuable information on soil and plant processes taking place in the irrigated field, and on management methods suitable for effluent irrigation.

The authors greatly appreciate the help of Mrs. Shulamit Gordon for her skillful and patient typing of the manuscript.

We would also like to express our appreciation to our colleagues in the Institute of Soils and Water, ARO, The Volcani Center, Israel for their constructive suggestions during the preparation of the manuscript.

Summer 1990

A. Feigin
I. Ravina
J. Shalhevet

Contents

Chapter 1 Introduction

Irrigation with sewage effluent provides crops with water and nutrients. At the same time it provides a convenient means of sewage disposal through land treatment, preventing potential health and environmental hazards caused by the uncontrolled flow of wastewater. Under arid and semi-arid environmental conditions, in which irrigation is a prerequisite for agricultural development, the importance of effluent as a source of irrigation water is as great as the need for its disposal. Effluent becomes exceptionally valuable where other sources of water are scarce, and its utilization for irrigation frees high quality water for other uses. Therefore, there is considerable interest in the use of sewage effluent in dry regions, such as parts of southern and western USA, Australia, Mexico, and Israel. Land treatment methods for disposal purposes are employed under arid as well as humid climates (Bouwer and Chaney 1974; Pound et al. 1983; Page et al. 1983).

According to available records, irrigation with sewage effluent has been practiced for centuries (Pound and Crites 1973b; Shuval et al. 1986). Some of those treatment plants presently used for recycling effluents have been active for decades; for instance, the Werribee farm, located near Melbourne, Australia, has been successfully operated since 1897 (McPherson 1979). Presently there is a growing interest in effluent usage for irrigation, resulting in the construction of new plants, as shown by many new publications (e.g., Page et al. 1983; Cowan and Johnson 1984; FAO 1985; Pettygrove and Asano 1985; Ali 1987; Arar 1987). Page et al. (1983) reported that more than 1000 effluent irrigation systems are being operated now in the USA.

Valuable information on effluent irrigation has been gained in Israel, where shortage of water stimulated research on different aspects of the subject (see Noy and Feinmesser 1977; Shelef 1977; Feigin et al. 1978).

Treated sewage effluent as a source of irrigation water may differ from the regular water supply in the following aspects:

1. Through municipal and certain industrial usage, biodegradable organic matter, as expressed in the BOD and COD indices, increases. The conventional reclamation processes result in a reduction in the organic matter content to any desired level, depending on the intensity of the treatment.

   The effluent may also contain trace organic toxic substances such as pesticides, but available information does not indicate an occurrence of health problems related to the presence of such compounds in municipal wastewater.
2. Municipal use invariably causes an increase in the inorganic soluble salt content of the water. The principle ions picked up are sodium, chloride and

bicarbonates. These ions bring about an increase in the total salt content (salinity) and in the sodicity of the water. Unlike organic constituents, the inorganic salts are not normally removed during the reclamation process, except for some precipitation of carbonates.

3. Effluents contain different levels of plant macronutrient elements, especially nitrogen and phosphorus. The presence of a considerable level of macronutrients in the soil should be taken into consideration as a source of plant nutrients as well as groundwater contaminants.
4. The concentration of trace elements may also increase, mainly from specific industrial wastewater. Certain trace elements required by plants (B, Cu, Mn, Mo and Zn) may be present in excessive levels; other elements, such as Cd, Pb and Hg, may also contribute to the toxicity hazard of the trace elements.
5. Pathogenic microorganisms (bacteria and viruses) are present, although their concentration in effluent is greatly reduced during the normal reclamation process. Parasites (protozoa and helminths) may also be present in sewage water.

Irrigation management using reclaimed effluent should take account of the special composition of the water described above, in addition to other considerations which apply to all irrigation water. These include water consumptive use (evapotranspiration), the choice of irrigation method, fertilization, drainage and other management practices.

Chapter 2 Sources, Treatment Processes and Uses of Sewage Effluent

2.1 Sources of Sewage Water

Wastewater is commonly classified as domestic, industrial or agricultural according to its source.

Sewage water originates mainly from domestic sources and contains varying amounts of industrial wastewater and occasionally rainwater. The quantity of sewage water produced is about 80% of the total domestic water consumption and varies between 100 and 300 l capita^{-1} day^{-1}, depending on the standard of living, climate, and season of the year. Industrial water use in industrialized countries is 6–10% of the annual demand and is predicted to reach 20–25% by the end of the century.

Information on wastewater flows and flow-rate fluctuations, which is required for designing treatment facilities, can be found in wastewater engineering textbooks (e.g., Fair et al. 1971; Bond and Straub 1974; Metcalf & Eddy Inc. 1979; Montgomery 1985).

A separate system of sewers for industrial wastewater is generally recommended for communities located in industrial areas, since they often contain substances deleterious to biological systems and may thus require chemical pretreatment. Some industrial wastewater contains toxic pollutants, which are usually very costly to remove. In such cases, it is preferable to treat the water in the plant before discharging it into the main sewer system.

Water reuse is an efficient means for some industries to meet the stringent environmental control regulations of wastewater disposal. However, internal recycling is impractical for industries using large volumes of water and different disposal methods are required. Irrigation provides one of the most practical and dependable means for reusing large volumes of treated sewage effluent.

Although agriculture is by far the largest water consumer, accounting for 70–80% of the annual water demand in many countries, it is not a major source of wastewater. More than 95% of agricultural consumptive use is for irrigation; most of the irrigation water evapotranspires to the atmosphere. Yet agricultural runoff, carrying fertilizers and pesticides, is a major cause of eutrophication of lakes and pollution of surface water and groundwater. However, the problem of irrigation return-flow management, increasingly important in connection with agricultural wastewater, is beyond the scope of this book. The daily quantity of wastewater produced by livestock and poultry facilities, consuming about 3% of the agricultural water demand, is estimated at 1 l per chicken, 25–30 l per sheep or swine, 40–50 l per head of cattle, and 100–150 l per head of dairy cow. Most animal wastewater is produced in confinement and can therefore be collected and treated on the farm for disposal or reuse.

2.2 Composition of Raw Sewage

Sewage effluent comprises 99.9% water and 0.1% organic and inorganic solids in settleable, suspended and soluble forms. Raw sewage water contains many microorganisms (bacteria, viruses, and parasitic protozoa), which may be pathogenic, and parasitic worms. Most of these are eliminated during the biological treatment processes and disinfection of the reclaimed effluent. *Coliform* or *fecal coliform* counts serve as standard indicators of microorganism contamination of sewage water. Typical raw domestic sewage contains 10^7–10^9 coliforms per 100 ml. Biological treatment and disinfection drastically reduce the number of fecal coli to the acceptable level for irrigation water (see Sect. 2.3).

Untreated domestic sewage contains 50–150 mg/l *grease* and *oils*, and the inclusion of industrial wastewater, especially from the food industry, may considerably increase the oil and fat levels. However, more than 65% of the grease and oils are removed by the flotation methods used in primary treatment. *Grit* and *settleable solids*, amounting to 5–25 mg/l in raw domestic wastewater, are completely removed during primary treatment.

Typical contents of organic and inorganic suspended solids range from 100–350 mg/l in raw municipal wastewater (Table 2.1), and the average pickup is estimated at 105 g capita^{-1} day^{-1}. The suspended solids content is an important parameter in evaluating the suitability of sewage effluent for irrigation, since these solids may clog both the soil pores and components of the water distribution system, especially small orifices. However, most of the suspended particles are settled in the treatment ponds and their concentrations in reclaimed effluent are generally low and comparable with those in fresh irrigation water.

Wastewater is a complex mixture of organic and inorganic compounds defying complete chemical analysis. As a result, a number of empirical methods have been devised for evaluating the concentration of organic contaminants in wastewater which do not require complete chemical analysis. The empirical method commonly used is the *biochemical oxygen demand* (BOD), which is the quantity of oxygen required for microbial degradation of the organic compounds in the wastewater at 20 °C. The standard BOD test carried out in the laboratory for 5 days (BOD_5) represents 70–80% of the total BOD (Idelovitch and Michail 1981). Another method is the *chemical oxygen demand* (COD), which is the quantity of oxygen necessary to oxidize the organic matter, and is usually determined by dichromate oxidation and generally produces higher values than BOD_5; it indicates the presence of persistent organic matter and reducing compounds such as ferrous ions, sulfite, sulfide and nitrite. Measurement of the *total organic carbon* (TOC) is also used for assessing organic contaminants. Higher values of these parameters (usually expressed in mg/l) indicate higher concentrations of contaminants, which may cause the development of anaerobic processes and the consequent offensive odors associated with sewage. Domestic

Table 2.1 Typical composition of raw municipal sewage (Pound and Crites 1973[a]; Bond and Straub 1974; Thomas and Law 1977; Idelovitch 1978; Asano et al. 1985)[a]

Constituent	Concentration, mg/l[b]		
	High	Medium	Low
Solids			
Total	1300	700	200
Dissolved	1000	500	260
Suspended	350	220	100
BOD_5	350	200	100
COD	1000	500	250
TOC	290	160	80
Nitrogen			
Total	85	40	20
Ammonium	50	25	10
Organic	35	15	5
Nitrate	1.5	0.2	0
Phosphorus	36	10	4
Chlorides	650	150	10
Calcium + magnesium	150	80	25
Sodium	460	120	10
Potassium	25	10	5
Alkalinity (as calcium carbonate)	400	200	50
Grease	150	100	35
pH	8.0	7.2	7.0

[a]Tables 2.1 to 2.5 give typical data on the chemical quality of raw sewage. However, due to the great variations in quality of the original water and other factors affecting the chemical properties of sewage water, a wide range of data is found in the literature. For example, the Cl^- levels commonly reported range between 10–750 mg/l, which stresses the need for adequate local information concerning the quality of wastewater used.
[b]Except for pH.

sewage is composed mainly of carbonaceous organic materials and nearly all of the oxygen demand is evaluated by the BOD_5 value. The average pickup of BOD_5 in domestic sewage is estimated at 95 g capita^{-1} day^{-1}. The COD/TOC ratio should theoretically be equal to the oxygen to carbon ratio (32:12 ± 2.67), but it varies within a wide range due to differences in properties of the organic matter. Table 2.1 gives the range of some quality parameters of raw municipal sewage water, but the presence of industrial wastes may change these values considerably. *Nutrients*, primarily nitrogen, phosphorus, and potassium, beneficial to agriculture, are also added to the supply water as a result of domestic water use.

Chemical characteristics of sewage water vary with the source of the potable water supply, the sewage system, the season and the nature of industrial discharge into the system. A major concern in using sewage effluent for irrigation is the presence of high concentrations of *hazardous constituents*, such as trace elements, stable organics and complex synthetic micropollutants. Some of these, e.g., zinc, copper and nickel, can be harmful to plants (phytotoxic) at excessive levels. However, the concentrations in most wastewater are well below the toxicity level for all crops, and phytotoxicity may occur only as a result of long-term accumulation in the soil. Heavy metals, such as cadmium, chromium and lead, can be taken up by plants resulting in toxic concentration in the food chain or pollute groundwater and surface water by deep percolation or runoff. Many of the trace elements present in wastewater originate from industrial effluent which, as stressed above, should be treated before being discharged into a municipal sewer. However, many of the trace elements in raw sewage are effectively removed into the sludge generated by primary sedimentation and secondary clarification. Trace element concentrations found in raw and untreated municipal wastewater are given in Table 2.2.

Boron, usually found in very low concentrations in fresh water, is added to sewage water as a result of soap and detergent use. Although the incorporation of boron into laundry powders has been restricted, wastewater from industrial laundering, hospitals or textile factories contain large amounts of detergents and phenols. Boron is desired in very small quantities for plant growth, but becomes toxic to many plants at slightly higher levels. Concentrations in excess of 0.5 mg/l in the soil solution are of concern and values higher than 5 mg/l have been reported for some effluents. It is estimated that 0.1–0.4 g B capita^{-1} day^{-1} is added to raw municipal sewage water.

Trace organics, some of which are known to be carcinogenic, are picked up in sewage water after usage of various insecticides. Although their concentrations are very low (μg/l range), many are resistant to degradation and thus may remain in the treated effluent, causing concern as potential pollutants of groundwater and surface water. It is generally believed, however, that these compounds are adsorbed by the soil matrix.

The principal criteria for assessing the suitability of sewage effluent for irrigation are the *total dissolved salt* content (TDS) and *sodium, bicarbonate* and *chloride* concentrations (Sect. 3.1). The salt concentration of water, usually measured by the *electrical conductivity* (EC) of the water at 25 °C, in dS/m, is an indicator of salinity hazard. The ratio of the amount of sodium to that of calcium plus magnesium in the water, known as the *sodium adsorption ratio* (SAR), and carbonate and bicarbonate contents are related to low soil permeability hazard. Chloride is toxic to certain crops such as citrus. All of these values are higher in sewage water than in fresh supply water and are not reduced by conventional sewage treatment. Some typical values of salt and mineral pickup from domstic water usage are given in Table 2.3.

Industrial wastewater, e.g., saline wastewater from pulp and paper manufacturers, may considerably increase the salinity of sewage water. Seasonal variations in salt pickup are also significant, especially in areas with specific industries, such as olive and pickle processing plants which discharge processing brine.

Table 2.2 Concentration of trace elements in raw and treated municipal effluents and the permissible level in irrigation water[a] and upper limit for drinking water for livestock[b]

Element	Raw wastewater		Primary effluent		Secondary effluent		Water quality criteria for irrigation[c]		Upper limit for drinking water (livestock)
	Range	Median	Range	Median	Range	Median	Long-term	Short-term[d]	
					mg/l				
Aluminium (Al)	—	—	—	—	—	—	5.0	—	5.0
Arsenic (As)	< 0.0003–1.9	0.085	< 0.005–0.03	< 0.005	< 0.005–0.023	< 0.005	0.1	10.0	0.2
Beryllium (Be)	—	—	—	—	—	—	0.1	—	0.1
Boron (B)	< 0.123–20.0	—	< 0.01–2.5	1.0	< 0.1–2.5	0.7	0.75	2.0	5.0
Cadmium (Cd)	< 0.0012–2.1	0.024	< 0.02–6.4	< 0.02	< 0.005–0.15	< 0.005	0.01	0.05	0.05
Chromium (Cr)	< 0.0008–83.3	0.400	< 0.05–6.8	< 0.05	< 0.005–1.2	0.02	0.1	20.0	1.0
Cobalt (Co)	—	—	—	—	—	—	0.05	—	1.0
Copper (Cu)	< 0.0001–36.5	0.420	< 0.02–5.9	0.10	< 0.006–1.3	0.04	0.20	5.0	0.5
Fluorine (F)	—	—	—	—	—	—	1.0	—	2.0
Iron (Fe)	—	—	—	—	—	—	5.0	—	Not needed
Lead (Pb)	0.001–11.6	0.120	< 0.02–6.0	< 0.2	0.003–0.35	0.008	5.0	20.0	0.1
Lithium (Li)	—	—	—	—	—	—	2.5	—	—
Manganese (Mn)	—	—	—	—	—	—	0.2	—	0.05
Mercury (Hg)	< 0.0001–3.0	0.110	0.0001–0.125	0.0009	< 0.0002–0.001	0.0005	—	—	0.01
Molybdenum (Mo)	< 0.0011–0.9	—	< 0.001–0.02	0.007	0.001–0.0018	0.007	0.01	0.05	—
Nickel (Ni)	0.002–111.4	0.230	< 0.1–1.5	< 0.1	0.003–0.6	0.004	0.2	2.0	—
Selenium (Se)	< 0.002–10.0	0.041	< 0.005–0.02	< 0.005	< 0.005–0.02	< 0.005	0.02	0.05	0.05
Vanadium (V)	—	—	—	—	—	—	0.1	—	0.1
Zinc (Zn)	< 0.001–28.7	0.52	< 0.02–2.0	0.12	0.004–1.2	0.04	2.0	10.0	24.0

[a] Chang and Page (1983); Page and Chang (1985).
[b] Ayers and Westcot (1985).
[c] The maximum concentration is based on a water application rate of 1200 mm/yr. In cases of higher rates, the maximum concentration should be adjusted accordingly.
[d] For use on fine-textured soils.

Table 2.3 Inorganic constituents added to effluents through domestic use[a]

Constituent	Range of increment mg/l
Total dissolved solids	150–400
Sodium	40–70
Potassium	7–15
Calcium	15–40
Magnesium	15–40
Chloride	20–50
Carbonate	0–10
Bicarbonate	50–100
Sulfate	15–30
Silica	2–10
Alkalinity (as calcium carbonate)	100–150
Boron	0.1–0.4
Phosphate	5–15
Ammonium	15–40

[a]Metcalf & Eddy Inc. (1979); Asano et al. (1985).

Na and Cl ion contents can be particularly high in wastewater from communities that have hard water and employ water softeners using salt brine for regeneration. Salinity parameters, in particular Na, Cl and HCO_3 ion contents, which affect crops and soil, are not eliminated by conventional wastewater treatment processes, and are a concern to the agriculturalist regardless of the source of the irrigation water. In general, the technical capability exists for removing any pollutant from wastewater. However, the necessity to do so should be examined very carefully because the marginal cost of unconventional treatment is exceedingly high.

2.3 Principal Treatment Methods

Untreated wastewater is a hazard to both public health and to the environment. The enhanced wastewater production stemming from increasing population and industrialization, together with intensified environmental awareness, have resulted in the enforcement of wastewater treatment by stringent water pollution laws and regulations in most developed countries.

Wastewater treatment processes are conventionally classified into *primary*, *secondary*, *tertiary* and *quaternary* treatments (Shuval 1977). The higher the level of treatment, the higher the quality of the effluent and consequently, its value. However, a more advanced treatment involves higher costs for construction, maintenance, operation and energy consumption, and the costs rise steeply with each advance in treatment level. The cost of tertiary treatment, including nitrification, denitrification, filtration, carbon sorption and disinfection, can be threefold or more than of simple secondary treatment using oxidation ponds, and about eightfold that of anaerobic facultative ponds (Shelef 1977). Thus, the necessity for advanced treatment should be carefully assessed according to the intended reuse.

Primary treatment removes coarse organic and inorganic solids, grease and oils from wastewater by screening, settling and flotation processes, and also conditions wastewater through neutralization and/or equalization before either disposal or discharge to a secondary treatment facility.

Secondary treatment involves both aerobic and anaerobic biological processes, in which complex organic matter in the wastewater is decomposed or oxidized by simple microorganisms. Conventional secondary treatments utilize activated sludge, trickling filters, aerated lagoons and oxidation ponds.

Tertiary treatment, employing chlorination, microscreening or filtration, coagulation, precipitation and activated carbon adsorption, further removes suspended particles, BOD, nutrients, eutrophication factors and turbidity, and virtually eliminates residual pathogens.

Quaternary or *advanced treatment* aims at upgrading the effluent to the level of fresh potable water and employs techniques such as ultrafiltration, ion exchange, reverse osmosis, electrodialysis or distillation, by which any undesirable constituent, including excess salinity, can be removed from the water.

The level of treatment required depends mainly upon the intended use or the environmental regulations for discharge of the effluent. In some cases, water quality criteria are unnecessarily stringent for agricultural irrigation use and may restrict effluent reclamation projects. The percolation of wastewater effluent through the soil profile, especially the active root zone, is a very effective renovation process, since chemical, biochemical and adsorption reactions remove or inactivate most pollutants and nutrients are taken up by plants.

Sewage water intended for irrigation is treated mainly in order to protect the public both from consuming contaminated crops and from direct exposure to the applied effluent. Additional reasons for treatment are the prevention of nuisance conditions, operational problems in the irrigation system and adverse effects on soil and crops. Public health requirements involved in effluent irrigation are discussed in Sect. 3.9.

2.3.1 Considerations in Selecting Treatment Level

In addition to the required effluent quality, selection of wastewater treatment processes depends on the following criteria:

1. Wastewater characteristics, e.g., BOD, pH, amount of suspended solids and presence of toxic material
2. Economic factors, such as cost and availability of land (certain biological processes, e.g., stabilization or oxidation ponds, are only economically feasible if low-cost land is available)
3. Specific public requirements, e.g., complete elimination of offensive odors
4. Considerations of a possible future upgrading of the effluent

A summary of standards pertaining to renovated wastewater for agricultural use in different countries was published by the WHO (1973).

Less stringent standards have been recently suggested for agricultural wastewater use (Engleberg Report 1985; Pescod 1986; Shuval et al. 1986; Shuval 1987; WHO Scientific Group 1989; see Sect. 3.9).

2.3.2 Primary Treatment Methods

Primary (mechanical) treatment, intended to remove grit and grease, is usually employed as a pretreatment for biological processes. The raw sewage water passes through bar racks and/or comminutors into grit chambers, where sand and other heavy particles are settled out. The effluent is then discharged into primary clarifiers or settling basins, or into flotation chambers where it is detained (usually for several hours) to allow the lighter and smaller solids to settle and grease and oils to rise. Grease and oils are skimmed off mechanically and the heavier solids are scraped off the bottom and delivered to a sludge handling and disposal facility.

Primary-treated effluent has a gray milky color which does not affect its use for irrigation but does indicate the presence of a high BOD which, under poor management, produces offensive odors. Primary effluent still contains 50–75% of the BOD, 50–65% of the COD and 30–50% of the total suspended solids found in raw sewage water (Asano et al. 1985). Although 50–70% of the total suspended solids of raw wastewater are removed by primary treatment, concentrations >100 mg/l are found in some primary-treated effluents (Table 2.4). Of the heavy metals present in the incoming wastewater, 40–50% are removed into the sludge accumulated at the bottom of the settling basins. Primary-treated effluent is

Table 2.4 Typical properties of primary effluents (Thomas and Law 1977)

Constituent	Concentration Range	Concentration Median
	mg/l	
Solids		
Total dissolved	200–1500	500
Total suspended	50–150	100
BOD	65–200	135
COD	150–750	335
Nitrogen		
Total	10–60	40
Free ammonia	7–40	30
Nitrate	—	<0.1
Phosphorus, total	5–17	8

acceptable for irrigation in isolated locations with restricted public access and is limited to crops not intended for direct human consumption. Primary treatment may be the only requirement if storage with a detention time longer than 2 weeks and impoundment depth <4 m is considered, since storage reservoirs act as stabilization ponds in which biological processes substantially reduce pathogenic microorganisms, BOD and suspended solids.

2.3.3 Secondary Treatment Methods

Secondary (biological) treatment is the more general requirement in the case of wastewater intended for irrigation. The simplest secondary treatment is provided by *oxidation ponds*, in which oxygen is produced by natural surface aeration and by algae through photosynthesis, involving consumption of nutrients from the wastewater. The oxygen is used by bacteria for decomposition of organic matter and the nutrients released are consumed by the algae, thus forming a symbiotic cycle. This type of treatment is suitable for rural communities and small or medium-sized urban municipalities (up to about 40 000 inhabitants), in regions of hot to moderate climate and no land shortage.

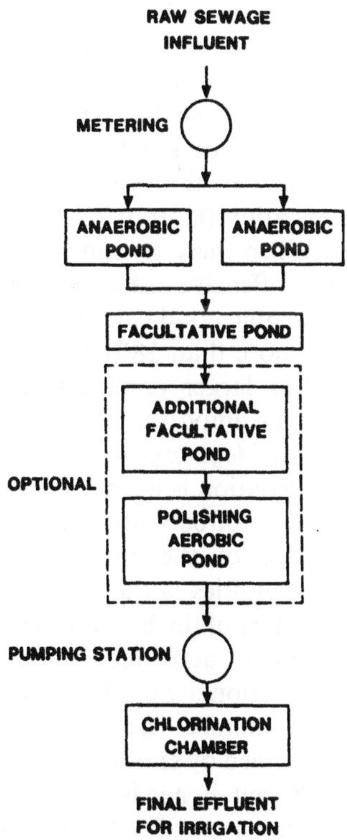

Fig. 2.1 Flow diagram of oxidation pond treatment

The oxidation pond system illustrated schematically in Fig. 2.1 is widely used in Israel. Raw sewage water is conveyed into small anaerobic primary ponds, where grit and sludge settle at the bottom. The anaerobic ponds, usually constructed in a parallel pair to enable alternate removal of sludge, are loaded with 500–2500 kg BOD_5 ha^{-1} day^{-1} (the common load in Israel is 1500–2000 kg BOD_5 ha^{-1} day^{-1}). The depth of the anaerobic pond ranges from 1.5–4.0 m (in Israel, a depth of 2.5 m and detention time of 1 day are usual). The rate of anaerobic biodegradation increases very steeply with increasing temperatures in the 14–40 °C range, while at temperatures below 12 °C the process virtually ceases.

From the anaerobic ponds, the primary effluent, with a 50–60% reduced BOD_5, flows into a successive series of facultative-oxidation ponds, in which upper aerobic zones are formed above the deeper anaerobic layers. The degree of aerobiosis (depth of aerobic layer), which generally increases in the successive ponds, depends on oxygenation and this in turn depends on solar radiation. The last pond in the series may become completely aerobic and is then referred to as a polishing or maturation pond.

In small rural communities (500–1500 inhabitants) in Israel, successful treatment is achieved with a single oxidation pond of shallow depth (1.2–1.5 m assures aerobic conditions even during winter) following the anaerobic ponds. The organic loading on the oxidation pond is about 150 kg BOD_5 ha^{-1} day^{-1}, with a detention period of 7–10 days. In many cases, it is advantageous to discharge the oxidized effluent into an operative reservoir, in which an additional detention of between 2 and 4 days upgrades the effluent quality. A gravel filter or fine screen at the outlet of the pond, followed by chlorination with a dosage of about 20 mg/l and contact time of about 1 h, can raise the quality of the treated effluent to that required for unrestricted crop irrigation and advanced trickle irrigation (Kott 1970).

When these treatment systems are well operated, they produce very little, if any, offensive odor. However, for complete elimination of such odors, the anearobic pond can be omitted by limiting the organic load on the first pond to not more than 180 kg BOD_5 ha^{-1} day^{-1}.

For larger communities, and especially when land cost and availability become important factors, more sophisticated treatment methods are used. *Aerated lagoons*, in which wastewater flows through ponds 1.5–4.0 m deep and oxygenation is accomplished by aeration units, are suitable for communities of 30 000–100 000 inhabitants, and require less land than simple oxidation ponds.

One of the most common secondary treatments employed in larger communities is the *activated sludge* process, a continuous operation based on recycling of the biological sludge (see the flow diagram in Fig. 2.2). The so-called extended activated sludge process uses a much larger reactor than is conventionally used, so that both the detention time and aeration are increased; this minimizes the amount of excess sludge, since all degradable sludge formed is consumed by endogenous respiration. The purification capacity of this method is high, but it needs a careful operational control. The area required for such systems is small (about $50\,m^2(10^3\,m^3\,day^{-1})^{-1}$).

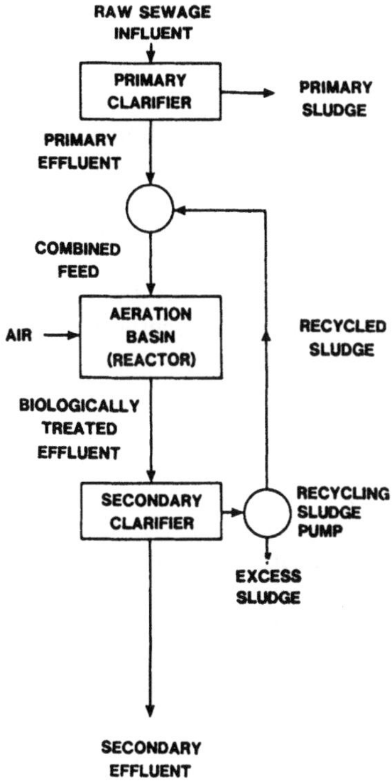

Fig. 2.2 Flow diagram of activated sludge process

Another common secondary treatment, suitable for low-flow-rate wastewater, uses *trickling filters*, which are packed with coarse media covered with biological slime, through which wastewater is percolated. The 0.1–2 mm-thick slime layer consists of aerobic and anaerobic sublayers. A combination of trickling filters followed by activated sludge is found advantageous for large cities.

Properly treated secondary effluent has very little odor, is clear and almost colorless. Clarified secondary effluent usually contains less than 30 mg/l BOD_5 and 20–30 mg/l suspended solids. Although more than 99.9% of pathogenic microorganisms found in raw sewage water are removed during treatment, the remaining coliform concentration is still very high and may exceed 10^5/100 ml. The properties of secondary effluent are further discussed in Sect. 2.8.

2.3.4 Disinfection

Disinfection is an effective means to further reduce bacteria and viruses in secondary effluent. Chlorine is the most commonly used disinfectant and, with sufficient amounts and contact time, it is possible to meet bacteriological discharge requirements and greatly reduce health risks. Applying 10–15 mg/l free chlorine for 5 min at pH 6.5 eliminates 99% of the spores of fungal plant pathogens. In Israel, secondary effluent from oxidation ponds is treated with 20 mg/l chlorine for about 60 min, which virtually removes bacteria and viruses (Kott 1970) and enables such effluent to be used for irrigation of crops for direct human consumption. No deleterious effect on soil biological activity or crop growth due to irrigation with chlorinated effluent has been reported. Yet, residual chlorine may involve leaf-tip burn and damage to sensitive crops if the free available chlorine (Cl_2) exceeds 0.05 mg/l. Furthermore, toxic effects of chlorinated organics in relation to ground water pollution are also considered (Pettygrove and Asano 1985). Ozone (O_3), produced at relevant treatment plants by passing dry air between two high-potential electrodes, is also used for disinfection. This method is popular in Europe. Its main advantage is the absence of hazardous pollutants.

2.3.5 Advanced Wastewater Treatment

Tertiary treatment is not necessary for reuse of treated sewage effluent for agricultural irrigation. However, if required, further removal of suspended solids, BOD and nutrients remaining in the secondary clarified effluent can be achieved by flocculation, which utilizes a coagulant such as alum, polyelectrolyte or lime, followed by a gravel and sand filter or any other graded filter to remove the flocs. This treatment is effective for removing bacteria and viruses which collect in the flocs, as well as most of the suspended solids, BOD, nutrients, heavy metals and turbidity. If lime is used, the lime-clarified effluent, with pH of approx. 11.5, must be neutralized with CO_2. In a plant located at the Greater Tel Aviv area in Israel, sewage water designated for groundwater recharge is neutralized by dissolving

Table 2.5 Quality of effluents resulting from advanced wastewater treatment in California (Asano et al. 1985) and Israel (Idelovitch 1978)

Parameter[a]	Six locations in California (Range)	Tel-Aviv, Israel: Tertiary treatment (Range)	Tel-Aviv, Israel: Tertiary treatment and groundwater recharge
BOD	2–9	3	<1
COD	—	63–85	20
Suspended solids	1–5	10–13	0
NH_3-N	0.1–16.6	3–10	0.1
NO_3-N	0.4–21.3	—	—
Total-N	—	6–13	5
Total-P	—	0.2–1.0	0.1
B	0.6–1.33	0.27–0.29	0.25
Alkalinity	150–256	110–122	100
EC (dS/m)	1.01–1.44	—	—
Coliforms MPN/100 ml	2–4	—	—

[a] Values given in mg/l, except as noted.

atmospheric CO_2 into the water in the polishing ponds to lower the pH. During the high pH period, ammonium is converted to free ammonia which is slowly released into the atmosphere. Within 2 weeks in summer, almost all NH_4^+ is thus removed, and in winter about 80% (Idelovitch 1978). The high pH environment which prevails prior to neutralization kills bacteria and viruses so that chlorination is unnecessary before recharging the groundwater, and the potential hazard of halogenated organics is reduced.

Activated carbon sorption is used to remove persistent organics from treated effluent. Activated carbons are prepared carbonaceous materials such as, wood, lignite, coal and nut shells, by a thermal activation process that yields a very porous material with a large specific surface area (as high as $1000\,m^2/g$) onto which solute molecules are adsorbed. Due to high production costs, this process is not widely used. However, as the demand for new water resources increases and effluent quality requirements become more stringent, the use of both activated carbon sorption and of more sophisticated and costly quaternary treatment methods will increase.

Typical properties of tertiary treated effluent are shown in Table 2.5.

2.4 Reuse and Disposal of Sewage Effluent and Sludge

2.4.1 Reuse of Effluent

In the past, wastewater was disposed of in the easiest way available without due regard to nuisance conditions, health risks or environmental damage. Sewage was discharged directly to the sea, lakes or rivers, or spread on land, relying on renovation by natural processes. With increasing population density and industrialization, the assimilative capacity of the environment has become limited. To protect the aquatic environment, legislation on the quality of discharged sewage effluent was introduced, forcing wastewater treatment to meet disposal requirements. Nevertheless, the method of final disposal has not been significantly changed, and only a very small proportion of wastewater is presently reclaimed for reuse.

Effluent that is discharged into land and inland water bodies is indirectly used, but most treated sewage effluent is discharged into non useable saline water bodies. The US Water Resources Council estimated that in 1970 only 3% of the reclaimed municipal wastewater was reused for planned purposes and the use of sewage effluent in 1982 in Australia was about 11% of the total annual flow (Strom 1984). The recent increase in water conservation awareness has increased water reuse efforts, and sewage effluent reuse has become a fundamental tenet of most reclamation projects. In Israel, 67% of the potential quantity of effluents are used for irrigation or reclaimed by recharging the aquifer (Eitan 1989).

Potential uses of treated wastewater include the following: (1) cooling water in industry; (2) groundwater recharge for seawater intrusion control or for integration in the general water supply system; and (3) potable water for direct domestic use. However, the most reasonable and applicable reuse is for irrigation of agricultural crops, parks and other landscape areas.

The risk to public health is the greatest concern in water reuse. Thus, domestic use of reclaimed effluent requires not only sophisticated and costly treatment, but also rigorous quality control. Direct reuse is technically feasible, but, apart from economic considerations, psychological barriers must first be overcome before reclaimed effluent will be used universally on a large scale (Huang 1982).

A more common reuse of effluent produced by high-level wastewater treatment, including nitrate and phosphate removal, is groundwater recharge. After dilution and recovery, the effluent becomes part of the overall nonrestricted water resources. It is essential, however, that extensive quality monitoring and control be maintained in the system to ensure that the effluent does not adversely affect the quality of the general water supply. Groundwater recharge for salt-water intrusion control is much less demanding in this respect.

Reclaimed wastewater is used in industry, particularly as cooling water. Treatment for such usage is required to reduce development of slime and algae (by reducing BOD and nutrient content), to eliminate corrosion and scaling agents and to reduce health hazards. Although this involves relatively advanced treatment, the economic return of such reuse may be higher than from agricultural irrigation.

The level of treatment required to produce reclaimed wastewater adequate for irrigation is significantly lower than that needed for other reuses and, furthermore, irrigation can consume large volumes of effluent. Irrigation with sewage effluent (see Chap. 4) has been considered in the past mainly as a disposal operation for treatment plants and the effluent has not been regarded as a source of water. This evaluation will eventually be altered due to economic factors, i.e., when the cost of reclaimed wastewater becomes lower than that of alternative sources. Reclaimed wastewater is generally of lower quality than conventional irrigation water due to its higher salinity level. However, most secondary effluents contain nutrients, so that to maximize the benefits of using such water, appropriate irrigation and fertilization management should be imposed (see Chap. 5). This, as well as health and environmental concerns, necessitates better water and soil monitoring than is used with conventional water.

Agricultural water demands vary with the season, whereas wastewater supply is continuous, thus seasonal storage or some alternative means of disposal of excess effluent must also be considered (Sect. 2.7). The engineering and agronomic aspects associated with wastewater reclamation for irrigation are well developed; this type of water reuse has proved to be viable and effective.

2.4.2 Sludge Disposal

Most primary and secondary treatment processes yield *sludges* requiring suitable disposal, which presents one of the most difficult environmental problems (Loehr 1977; Eliott and Stevenson 1977; Page et al. 1983, 1987). As with all waste disposal, the basic requirement is that the operation will not adversely affect health or the environment.

Sludges resulting from sedimentation or floatation processes, usually referred to as primary sludge, have a solid concentration of 0.5–5%. Treatment is generally required before final disposal. One approach is to reduce the total amount of organic and volatile constituents in sludge liquid by subjecting it to aerobic or anaerobic digestion. Another method is to increase the sludge solid content to 15–50% by a sequence of thickening and dewatering processes.

Sludge disposal is particularly problematic in large wastewater treatment plants serving metropolitan areas, where the volume requiring disposal is very large. Ocean dumping, wherever possible, is the simplest and most convenient solution. Unprocessed or thickened sludge is pumped into deep water through pipes a few kilometers long.

Sludge incineration is another frequently employed disposal method. Dewatered concentrated sludge is introduced into the incinerator, where internal combustion is possible after the temperature is raised above ignition point by burning an auxiliary fuel. The combustion products are mainly carbon dioxide and sulfur dioxide, the emission of which is governed by air quality regulations, and ash, which must also be disposed of.

A commonly utilized approach involves various land disposal methods. Sludge from small treatment plants is often disposed of in lagoons, oxidation ponds similar to those used in wastewater treatment, sanitary landfills, septic tanks, sludge trenches or sludge drying beds. The USEPA has already established criteria that define acceptable disposal practices so that hazardous components do not leach through the soil, are not washed off the site in runoff, are not blown away, are not transmitted to the food chain, or otherwise threaten public health or èndanger sensitive land and biological resource.

Since sludges contain valuable nutrients, various utilization possibilities are being considered, including composting and processing for fertilizer products and animal feed. The most obvious utilization is land application for agricultural crop production, and in the USA, about 30% of municipal sludge production is already being used for land application.

The main concern is the content of heavy metals (Page 1974; CAST 1976; Chaney et al. 1987; Page et al. 1987), especially cadmium. The EPA has issued regulations governing sludge application to a depth of 1 m within the root zone of food crops where cadmium uptake is likely. (Trace element aspects are discussed in Sects. 3.6 and 5.4.)

For most agricultural uses, the sludge application rate will be limited by nitrogen requirements. In addition, since organic nitrogen is slowly released over a period of years, repeated applications, not even exceeding the annual nitrogen requirement level, could cause a buildup to toxic levels. Thus, regular soil and leaf analyses are essential to monitor the situation.

In the USA, sludge cannot be applied· directly to the land but must be pretreated by one of the following processes: (1) aerobic or anaerobic digestion; (2) composting; (3) drying; (4) lime stabilization or any other method that will significantly reduce pathogens. Public access to application sites must be limited for at least 12 months and grazing must be prevented for at least 1 month. If crops for human consumption are to be grown on the land within 18 months of application, the sludge must undergo additional processing to further reduce pathogens or it must be radioactively irradiated or pasteurized prior to application. Due to the high costs involved, these regulations may severely limit land disposal.

2.5 Land Treatment of Sewage Effluent: Approaches, Systems and Methods

Land application of organic wastes and wastewater has been practiced for many years as a treatment process and means of disposal. When properly designed, land application systems for sewage effluent are efficient. Irrigation of agricultural crops can be part of a land treatment system, or the renovated water can be discharged to groundwater or surface water for reuse.

The soil and plants act as a "living filter," absorbing and retaining pollutants and pathogenic organisms present in wastes and wastewaters. Land application of partially treated wastewater completes a sequence of treatments for reducing the level of microorganisms and various organic and inorganic compounds to acceptable levels for reuse of the renovated water. The main criticisms of land treatment methods are based on the following factors: (1) public health hazard, such as transmission of pathogenic bacteria and viruses to man and animals, contamination of groundwater by hazardous chemicals and propagation of those insects which are disease vectors; (2) long-term effects on soil detrimental to its physical and chemical properties; (3) shortage in land availability; and (4) public acceptance of products obtained from crops grown on soils irrigated with sewage effluent.

Soil, being a dynamic, living system, characterized by a large active surface and a multitude of physical, chemical and biological processes, reacts strongly with the constituents in sewage effluent. According to Miller (1974) the soil may effectively eliminate pathogenic bacteria and protozoa reaching the soil when secondary effluent or sewage sludge is applied. The survival time of pathogenic organisms in the soil depends on the type of organism and soil conditions, can vary from days to months and should be considered as a factor in the design of land application systems. Organic compounds applied with the wastewater decompose to CO_2, water and inorganic compounds. Inorganic constituents may be exchanged, adsorbed or precipitated, following chemical reactions which transform them into low solubility compounds, or they may be taken up by plants and consequently be partially removed from the flowing solution. The fate of materials added with the waste to the system depends on their physicochemical characteristics as well as on soil, vegetation, climate and management factors. These and other regulatory, economic and technical factors must be taken into account when selecting the site and the most suitable land treatment method and its design (Pound and Crites 1973a, b; Kidder 1976; Page et al. 1983; Pettygrove and Asano 1985).

Regulatory aspects involve laws, regulations and criteria concerning the protection of public health and preservation of streams, groundwater and environmental quality. Economic factors are the amount of revenue from the land, the renovated water and the irrigated crops. Technical factors include wastewater characteristics, land availability, soil type, climate, underground formations, topography, type of available equipment, wastewater flow rate and whether it is constant or seasonal.

The three basic approaches of land treatment of sewage effluent are (1) irrigation (a low rate process); (2) overland flow; and (3) infiltration-percolation (high rate processes); (Kidder 1976; Pound et al. 1983). Land treatment methods are site specific and many design possibilities are available. Planning of a land application system must be multidisciplinary, involving such fields as hydrology,

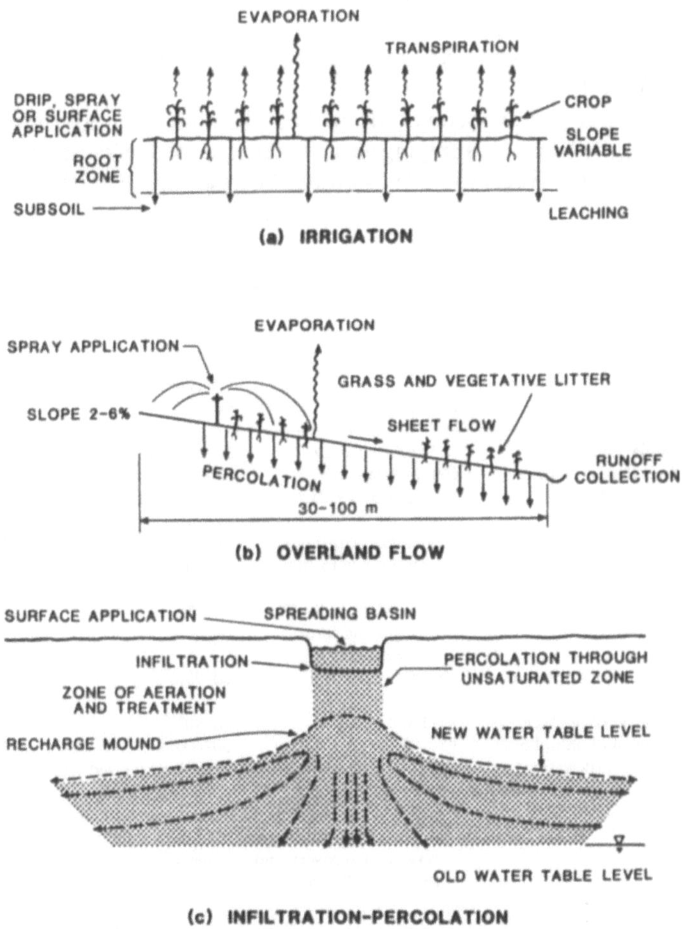

Fig. 2.3 a–c Land application approaches (After Pound and Crites 1973b)

soil science, agronomy, horticulture and environmental engineering. A process design manual for land treatment of municipal wastewater is available from the EPA (1981). Figure 2.3 shows schematically the main components of the three land treatment approaches and Table 2.6 presents the main comparative characteristics of the systems.

Irrigation with low application intensity (e.g., < 7.5 cm/week) is economical for agricultural crops, especially where the price of water is high. High-intensity irrigation (7.5–10cm/week) is used for grasses tolerant to low aeration where fresh water is abundant and the main objective of the effluent application is wastewater treatment. Much higher rates of effluent (usually secondary effluent) are added in infiltration-percolation systems to basins or by means of high-rate sprinklers in highly permeable soils. The percolated water can be reused through recharge of groundwater or recharge by deep, lateral-flow, surface water courses, or it can be recovered by underground drainage. Information regarding the processes and operational aspects of infiltration-percolation was obtained in the Flushing Meadows Project in Arizona (Bouwer 1970; Bouwer et al. 1974a, b; Bouwer and Chaney 1974).

Overland flow systems, also called filtration or spray runoff systems (Harlin 1979), are based on applying wastewater to sloped terraces or low permeable soils covered with vegetation. Wastewater flows along a 30–100-m downslope (2–6%)

Table 2.6 Comparative characteristics of low-rate (irrigation) and high-rate (overland flow and infiltration-percolation) systems (After Kidder 1976)

	Design approach		
Factor	Irrigation	Overland flow	Infiltration-percolation
Liquid loading rate	12.5–100 mm/week[a]	50–140 mm/week	100–3000 mm/week
Annual application	600–2400 mm/yr	2400–7200 mm/yr	5500–15,000 mm/yr
Land needed per 1000 m^3/d	15–60/ha plus buffer zone	5–15/ha plus buffer zone	0.2–7/ha plus buffer zone
Soils	Moderately permeable loamy sands to clay loams	Slowly permeable silt loams to clays	Rapidly permeable sandy loams to sands
Slopes	Cultivated crops: 0–6%. Forages and forest species: 0–15%	2–6%	less than 2%
Removal of suspended solids and BOD	90–99%	90–99%	90–99%
Removal of nitrogen	80–100%	70–90%	0–80%
Removal of phosphorus	95–100%	50–60%	70–95%
Fate of wastewater	Evapotranspiration and deep percolation for groundwater recharge, discharge into surface waters, or recovery and reuse; runoff controlled	Runoff maximized for recovery and reuse; relatively little evapotranspiration or deep percolation	Deep percolation maximized for groundwater recharge, recovery and reuse; runoff not allowed; negligible evapotranspiration

[a]Irrigation at 100 mm/week would be seasonal. A 2400 mm/yr application would average 60 mm/week over a 40-week irrigation period.

across grass or vegetative litter and is renovated by physical, biological and chemical processes. When used as a secondary treatment, overland flow is efficient in reducing BOD, suspended solids and N from primary effluent, but it is not so efficient in reducing P levels (Pound and Crites 1973b; Kidder 1976). It is generally more efficient under warm weather conditions than in cold zones (Jenkins and Martell 1979). As an advanced treatment, overland flow is less efficient than a low-rate, land-application irrigation system (Kemp et al. 1979).

2.6 Integrating Treated Sewage Effluent in the Overall Water Resources

It is clear that when available freshwater supplies are limited, the urban wastewater must be viewed as an agricultural resource. Israel has already reached the stage when conventional groundwater and surface water resources are fully utilized. Since sewage water must be treated before disposal in accordance with water pollution and environmental laws and regulations, it would be economically wasteful if the treated water were not utilized. In many locations where water is not yet limited and a choice exists between using fresh water or reclaimed wastewater, farmers are prejudiced against the reclaimed wastewater unless it is offered at a significantly lower price. It has been proven experimentally and in practice that treated sewage effluent can provide an additional source of irrigation water to be included in the overall water management program. This source is particularly valuable during drought years, when the agricultural water demand is higher than normal and the supply of fresh water for municipal and industrial uses cannot be drastically reduced. Consequently, agricultural water use must be cut to offset the deficiency. Utilization of effluent may also reduce the need for new freshwater development projects and the high cost of advanced treatment and disposal facilities. Thus, on a regional basis, the benefits from using reclaimed wastewater may even exceed the obvious advantages to the individual farmer.

Reuse of sewage effluent as an integral part of the non restricted water supply system is hindered by the high cost of the advanced treatment, quality control and monitoring that are required. It is likely that, in the near future, the use of different grades of water for different needs will be the more general approach for integrating reclaimed wastewater into the total water resources system; this will probably necessitate dual distribution systems.

2.7 Storage of Treated Effluent

To make full use of reclaimed wastewater, there is a need for effluent storage, on both an operational and seasonal basis. Storage may be necessary for effluents during periods of low water demand for their use during periods of peak irrigation demand in excess of supply. This peak period may be relatively short, as is the case of cotton, the crop most commonly irrigated with treated sewage effluent in Israel today. This crop requires irrigation for only 60–90 days per year. Direct groundwater recharge and consequent pumping is one possible storage method for reclaimed wastewater. Another is percolation through the soil and dilution in the aquifer, which also improves the quality of the effluent. Discharging into lakes and rivers is another possibility. However, unless these waters are restricted in its general use, very high levels of treatment and monitoring are required.

Seasonal impoundment is the more common way to store treated sewage effluent. The use of storage ponds and reservoirs provide additional time for resolving temporary quality problems; they enable further treatments to be carried out, such as sedimentation, and disinfection and the blending of effluents of lower quality with higher quality water.

Seasonal storage during rainy seasons is desirable and the storage pond may also be designed to store floodwater, tail water and runoff. In Israel, more than 200 such ponds are successfully operated, many of which are multipurpose with a volume ranging from several thousand to several 100000 m^3. In small communities, the seasonal pond can be part of the treatment sequence, also serving as a facultative-oxidation pond.

In more densely populated urban areas, it may be expedient to transfer surplus effluent that cannot be used locally, or that can be used elsewhere with higher returns, to agricultural regions employing a central operational-seasonal storage reservoir for subsequent distribution to the farmers. For example, an annual flow of approximately 130 million m^3 of effluent from Greater Tel Aviv has been designated for transfer to the Negev, about 140 km to the south, to irrigate 35000 ha of cotton and grain crops. Central storage reservoirs for such a scheme require capacities of several million m^3.

Most storage reservoirs for agricultural purposes are constructed with earth embankments, designed in accordance with the principles stipulated for small earthfill dam design by the US Bureau of Reclamation (1974). In most cases, a competent soil engineer must be consulted for proper soil analyses and structural design of foundations and embankments. An important consideration in the design of sewage effluent ponds is prevention of percolation to groundwater; this can be achieved by lining the ponds. The depth of a reservoir depends on whether it is constructed with embankments or dikes on level ground or by damming a watercourse or ravine. Typical depths for storage ponds in Israel are 4–7 m, and within this range they are generally well mixed.

Since treated sewage effluent still contains nutrients and BOD, oxidation processes continue in the storage reservoir and anaerobic conditions may develop, especially in the deeper layers. However, if the upper layer (1–2 m) contains sufficient dissolved oxygen, the reservoir will not produce an offensive odor. Maximum oxygen concentrations occur in the afternoon (oxygen saturation is 7–8 mg/l) and minimum concentrations at dawn. If the dissolved oxygen content is less than 0.5 mg/l, it may rapidly decrease to zero (especially in smaller ponds), and then the anaerobic decomposition may produce sulfides which, in addition to having a bad odor, are corrosive. The volatility of sulfides depends on pH, at $pH < 6$, all sulfides are volatile.

The conditions in storage ponds are ideal for algal bloom. The blooming of photosynthesizing green algae in the upper layer (about 1 m deep) has certain advantages. The oxygen production, together with the natural mixing of water layers, helps in maintaining aerobic conditions in the pond. The CO_2 consumption increases the pH of the water to >8, which causes further precipitation of heavy metals. Most undesirable are the blue algae which proliferate in early autumn; they not only consume oxygen, but are slimy and tend to conglomerate, which may cause severe clogging problems in the pumping, filtering and distribution systems.

Since algae float in water and tend to pile up as a result of wind action, it is advisable to locate the pumping station on the windward side of the pond. Another important consideration is the depth from which to pump the water. Since the upper layer is rich in algae and the deeper layers may be anaerobic and corrosive, it is advisable to pump from a depth of 0.5–1.5 m below the water surface. However, since the optimum time and depth of pumping are stongly dependent upon the specific conditions in the pond, the pumping characteristics vary for each location and with time. It is recommended to design the pump inlet structure so that the pumping depth may be altered when necessary.

Water in reservoirs is also rich in aquatic animals (zooplankton) such as Rotatoria, Copepoda, and Daphnia, which proliferate in the presence of algae. These fauna inhabit the whole reservoir and can reach several millimeters in length and, if not filtered, may cause severe clogging of small orifices in the irrigation system.

The parameters characterizing an effluent reservoir which are important for monitoring purposes are thus: (1) concentration of dissolved oxygen; (2) temperature; (3) pH; (4) clarity or turbidity; and (5) biological constituents.

2.8 Properties of Secondary Effluent

2.8.1 Introduction

Quality standards are usually designed to obtain maximum benefit from irrigation water. Besides crop yield and its quality, soil fertility, groundwater properties and environmental quality are affected by the quality of irrigation water. This is particularly true in the case of reclaimed wastewater.

The term "water quality" refers to the concentration of dissolved and particulate constituents in the water, which may have direct or indirect influence on its use. The present discussion will be limited to those constituents which may be detrimental to crop production, either through their immediate effect or by virtue of their long-term, gradual accumulation in the soil and groundwater. These include salts, nutrients, trace elements and organic compounds. Some attention will also be given to microorganisms which may not affect crops but may be pathogenic to humans and animals.

Table 2.7 Characteristics of secondary municipal effluents (Pound and Crites 1973a; Thomas and Law 1977; Idelovitch 1979; Asano et al. 1985)

Constituent	Range
	mg/l
Total solids	400–1200
Total suspended solids	10–100
Total dissolved solids	400–1100
pH	7.8–8.1
BOD	10–80
COD	30–160
Total nitrogen	10–50
Nitrate-nitrogen	0–10
Ammonium-nitrogen	1–40
Total phosphorus	6–17
Chlorides	40–200
Alkalinity (Calcium carbonate)	200–700
Sodium	50–250
Potassium	10–40
Calcium	20–120
Magnesium	10–50
Boron	0–1
SAR	4.5–7.9

[a] All units are in mg/l, excluding pH and SAR $(mmol/l)^{1/2}$.

Appropriate analytical methods suitable for determination of the relevant quality parameters can be found in standard soil manuals and water or wastewater determinations (e.g., Taras et al. 1971).

The great potential use of secondary effluent in crop production (in comparison to less progressively treated waste water) makes its physical, chemical and biological quality a subject of a special interest. Typical properties of secondary effluents are presented in Table 2.7.

2.8.2 Salinity and Sodicity of Effluents

Three major characteristics describe the salinity and sodicity hazard with regard to agricultural water use of effluent:

1. The total salt concentration in the water, which may cause increased soil salinity
2. The concentration of specific ions such as Cl and Na which may have a direct or indirect toxic effect on plants, including nutritional imbalance
3. The concentration of specific ions (mainly Na and HCO_3) which may result in the deterioration of soil structure and consequently in reduced permeability. This criterion is usually expressed as the sodium adsorption ratio (SAR)

Total Salt Concentration. Most effluent waters contain higher concentration of soluble salts than the supply water from which they were derived. Salts are added to the water through domestic and industrial use. The quantities of salt added vary among locations and it may range between 100 and 800 mg/l, yet for any one location it does not change greatly. By and large, the absolute quantities added through municipal use are not large (Table 2.3), yet they are sometimes sufficient to change water of acceptable quality into water of questionable quality from the agricultural point of view. In the normal reclamation process salts are not removed.

Chloride. The level of chlorides in secondary effluent is usually higher than that found in the original supply water. In certain cases, such as in wastewater of some food industries, the increase of Cl concentration is very high due to the addition of salt in the food processing. Standard sewage treatment does not remove Cl from effluent due to the high solubility of Cl compounds. The various desalination technologies for removing Cl from water are still very expensive. Dilution of sewage effluent with high quality water will reduce the Cl level proportionally.

The typical range of Cl levels in most secondary municipal effluent (see Table 2.7) is below that considered hazardous for most agricultural crops. Municipal use results in the addition of 20–75 mg/l Cl (Table 2.3). Although sensitive crops may suffer from a Cl concentration as low as 200 mg/l, other crops may tolerate higher Cl levels. A high Cl content in sewage effluent, as well as in other sources of water may affect groundwater quality. Yet very high Cl levels (approx. 1500 mg/l) were also reported for effluents probably originated from saline water (Ali 1987).

Sodium. Due to its great impact on both soil properties and crop yields (the sodicity hazard; see Sect. 3.1), determination of the Na level in irrigation water is essential. Typical Na^+ levels in sewage effluent range from 50–250 mg/l (Table 2.7), but higher or lower values, depending on the level in the original tap water supply, have also been measured.

Calcium and Magnesium. The levels of Ca^{2+} and Mg^{2+} in effluent are often similar to those found in fresh irrigation water. The presence of Ca^{2+} and Mg^{2+} in irrigation water lowers the SAR value; consequently, water quality is improved (see Sect. 3.1). The application of Ca^{2+} and Mg^{2+} to acid soils increases the pH to a range more favorable to plants. Furthermore, since Ca^{2+} and Mg^{2+} are essential nutrients for plant growth, their supply, under the deficiency conditions often developed in acid soils, is beneficial to crop yields.

Alkalinity. Both HCO_3 and CO_3 ions may increase the sodicity hazard of water by causing the precipitation of calcium carbonates. At pH levels below 8.4, bicarbonate is the predominating ion, while above this pH the carbonate ion predominates. The pH of sewage effluent is usually slightly alkaline, but is rarely above 8.4. The concentration of HCO_3 in sewage effluent is much higher than that in the original supply water. Alkalinity, determined as $CaCO_3$, is

200–700 mg/1 (Table 2.7). The level of HCO_3 in different sewage effluent used in Israel for irrigation is about twice as high as that of fresh irrigation water.

2.8.3 Macronutrients

Nitrogen. The quantity of N in typical raw municipal wastewater ranges between 20 and 100 mg/l, the corresponding values for primary effluent are 10–60 mg/l (Thomas and Law 1977) with a median of 40 mg/l, and the range of N in secondary effluent (see Table 2.7) is 10–40 mg N/l. Similar figures were given by Pound and Crites (1973a, b): 40 mg N/1 and 20 mg N/1 for untreated sewage and for a typical treated secondary effluent, respectively. Field studies have shown that concentrations of N in effluent applied to land range from 10–60 mg N/l. Quin and Woods (1978) found the concentration of N in treated sewage effluent used for irrigation of pasture to be 10.6–16.7 mg N/l. The ranges cited in numerous publications reporting N levels in treated sewage effluent used for irrigation are almost identical to those given above.

The level of N in the raw sewage water and the treatment method employed determine the level of N in the secondary effluent obtained. Since the N level in sewage effluent greatly influences crop yields and may affect the quality of underlying groundwater, it is essential to obtain adequate information on N concentration in specific treated effluent intended for irrigation use. Although the N concentrations given above apply to most sewage effluents used for irrigation, higher or lower levels are possible, and levels below 10 mg N/l are not rare, especially in tertiary effluent (Idelovitch et al. 1976). Moreover, seasonal fluctuations are normal and should also be taken into account.

Sewage effluent contains N in four different forms: organic, NH_4^+, NO_3^- and NO_2^- (Pound and Crites 1973a,b). The level of NO_2^- is rarely greater than 1.0 mg/l due to its easy oxidation to NO_3^- in the presence of oxygen. The concentration of NO_3^- in secondary municipal effluent is relatively low, e.g., about 7% of the total N in secondary municipal effluent used in a field experiment carried out in Minnesota (Clapp et al 1977) and about 2% in field experiments performed in Israel (Feigin et al. 1978, 1979). The treatment method affects the proportion of the different N components in the treated sewage effluent, e.g., when the sewage effluent treatment involves forced aeration, nitrification may take place and considerably increase the percentage of NO_3-N in the effluent. The actual range of NO_3-N concentrations in secondary effluent used for irrigation in different sites in the USA is 0–10 mg/1.

Most of the N present in secondary municipal effluent is in reduced forms, mainly NH_4-N and organic forms. Typically, 80% of the total N in municipal effluents is NH_4-N, although values as high as 90–95% have also been recorded (Lance 1972). The NH_4-N/total N percentage in effluent used in field experiments (Feigin et al. 1984) in Israel was 65–95%. A large proportion of the sewage effluent-N can appear as organic N (Hunter and Kotalik 1974). Common sanitary analyses divide wastewater organics (nitrogenous as well as other organic materials) into suspended matter and soluble matter. The organic material is classified into a material that can be sedimented out, colloidal fractions obtained by centrifugation and superspeed centrifugation, and a soluble fraction.

The percentage of proteins of total COD is in the range of 10–20% (Hunter and Kotalik 1974). Rebhun and Manka (1971) detected compounds in effluent (e.g., humic acids) that contain considerable quantities of N.

Phosphorus. The level of P in sewage effluent varies within a wide range. According to Pound and Crites (1973a; see also Thomas and Law 1977), the levels of total P in strong-, medium- and low-strength raw municipal wastewater are 20, 10 and 6 mg/l, respectively. According to Hunter and Kotalik (1974), the quantity of inorganic P added to the sewage effluent as a result of common domestic usage is 25 mg/l. A typical range of P level in primary effluent is 5–17 mg/l (with a median of 8 mg/l), and the corresponding range for secondary effluent is 6–17 mg/l (Pound and Crites 1973a); see Table 2.7. Ryden and Pratt (1980) conclude that in most cases the concentration of P in effluent is less than 20 mg/l. Ellis (1974) suggested 10 mg total P/l as typical for most municipal wastewater. The concentration of total P in municipal sewage effluent used in a series of field experiments in Israel was 16 mg/l (Dag et al. 1984), 5–17 mg/l (Feigin et al. 1979) and 6–25 mg/l (Feigin et al. 1984). Lower values of total P are found in sewage effluent produced by highly efficient or advanced technologies (Table 2.5). The expected values of total P in the secondary effluent of the Dan Region Pilot Plant (Israel) were 10–12 mg/l; but the tertiary effluent produced by this plant contained only 0.2–1.0 mg P/l (Idelovitch et al. 1976). Sewage effluent contains different forms of P: organic, condensed phosphates (pyro-, meta- and polyphosphates) and orthophosphates.

Organic P forms are the result of biological processes and condensed phosphates are added by disposal of laundry chemicals into sewage. Organic and condensed P degrade to orthophosphate in the soil, but degradation may also take place in the wastewater treatment (Murmann and Koutz 1972, cited by Ellis 1974). Other sources indicate that the predominant chemically active P in soils is the orthophosphate form (Ryden and Pratt 1980). The total quantity of P added to soil by irrigation with sewage effluent is often considerable; values within the range of standard fertilization levels are widespread. For instance, the quantity of P added to different crops in secondary effluent irrigation experiments in Israel were as follows: a Rhodes grass field irrigated with 400 mm received 100 kg P ha^{-1} yr^{-1}, and wheat to which 200 mm supplementary irrigation was applied, received about 50 kg P ha^{-1} yr^{-1} (Feigin et al. 1978, 1979, 1984). Similar P levels were added in other experiments in Israel.

Potassium. The level of K in secondary effluent ranges from 10–40 mg K/l. Irrigation with sewage effluent thus adds large quantities of K to agricultural soils, reducing the need for K fertilization. For example, a cotton field irrigated with about 500 mm sewage effluent containing an average of 28 mg K/l received about 140 kg K ha^{-1} yr^{-1}, which is equivalent to about half of the annual consumption of K by cotton. However, in soils with low levels of available K, and especially where K-sensitive crops are grown, application by effluent alone is inadequate and additional fertilization is essential.

2.8.4 Organic Compounds, COD and BOD

Fresh water does not contain organic matter, thus the organic substance detected in sewage effluent results from domestic or industrial use (e.g., food processing) or are produced by biological activity during the course of secondary treatment (Hunter and Kotalik 1974). The estimated contribution of organic C by domestic use (g capita^{-1} day^{-1}) is: feces 17; urine 5; dishwashing and food preparation 8; and personal and laundry 7 (Hunter 1971). The corresponding organic N quantities are: feces 1.5; urine 1.7; and dishwashing and food preparation, 0.2 g capita^{-1} day^{-1}. The sewage treatments greatly reduce the level of organic matter, as shown in Tables 2.1, 2.4 and 2.7. The BOD and COD values are commonly used for evaluating the concentration of organic matter in sewage effluent as well as its environmental impact (see Sect. 2.2).

Recently, UV spectrophotometry has been successfully used for evaluating organic matter in water and wastewater and has shown good correlation with other methods (Idelovitch and Michail 1981). The organic matter found in sewage effluent is present in different physical phases; for instance, the percentages of organic C in the effluent obtained from activated sludge and trickling filter plants was as follows: soluble 52 and 69%, fine suspended 6 and 9%, and coarse suspended 25 and 39% (Hunter and Kotalik 1974). As may be anticipated, the proportions of organic substances present in sewage effluent differ according to its origin and treatment method. Rebhun and Manka (1971) and Manka et al. (1974), identified fulvic, humic and hymatomelanic acids in

Table 2.8 Organic constituents of secondary effluent from Haifa, Israel (The values present COD range and percentage of constituent of total COD)

Constituent	Trickling filter[a]	Trickling filter[b]	Stabilization pond[b]	Domestic wastewater; extended aeration activated sludge[b]
COD range (mg/l)	185	158–181	120–153	106–167
Ether extractable	6.6–7.8	13.4	11.9	16.3
Anionic detergents	12.8–15.6	16.6	12.2	16.0
Carbohydrates	10.9–12.7	5.9	7.8	4.6
Tannins	1.4–2.2	1.3	2.1	1.0
Proteins	21.0–24.1	21.6	21.1	23.1
Fulvic acid	18.8–25.8	25.4	26.6	24.0
Humic acid	10.4–11.9	12.5	14.7	6.1
Hymatomelanic	7.4–7.5	7.7	6.7	4.8

[a]Range of three samples (Rebhun and Manka 1971).
[b]Manka et al. (1974).

effluent samples from Israel (Table 2.8). Other authors (see Hunter and Kotalik 1974) could not identify all the organic matter present in the effluent; they found that the identified part consisted of ether extractables, proteins, carbohydrates, tannins and lignins, and different organic acids, amino acids and surfactants (methylene blue active substances, MBAS). The use of biodegradable surfactants reduces the detrimental environmental effect of detergents.

2.8.5 Trace Elements and Trace Organics

Trace Elements. The level of trace elements (As, B, Ba, Cd, Co, Cr, Cu, Hg, Mn, Mo, Ni, Pb, Se, V and Zn) in treated sewage effluents (secondary effluent included) is determined by the chemical properties of the raw sewage from which these effluents were derived and the treatment method used.

Trace elements are universally present in water (Ayers and Westcot 1985), but their level in fresh water is negligible. Domestic use rarely yields large quantities of trace elements in the wastewater, whereas industrial processes usually result in enhanced levels. Secondary sewage treatment reduces the trace-element content through the settling of suspended solids. The concentration of trace elements can be lowered by 70–90% through secondary treatment (Bouwer and Chaney 1974), and to an even greater extent by tertiary treatment and certain advanced technologies decrease the trace-element levels to those found in the original freshwater.

Table 2.2 presents typical concentrations of trace elements in raw, primary and secondary effluents. This table, as well as other available data, indicates that many sewage effluents are suitable for long-term irrigation. The upper limit for trace element threshold for sewage effluent irrigation is based on the most sensitive crops. Page and Chang (1985) therefore concluded that through good irrigation management, use can be made of effluent with higher concentrations than those specified in the table.

Trace Organics. Industrial wastewater contains various chemicals, such as organics acids, carbohydrates, organic bases, polynuclear hydrocarbons, phenols and aliphatic and aromatic materials (Hunter 1971).

Special attention has been paid recently to certain organic compounds found in water and wastewater at trace levels (Grigoropoulos and Smith 1971). There are over 100 chemicals of concern in sludge and wastewater (Overcash 1983). Great effort has been made to identify and remove these trace organics from wastewater (e.g., Cooper 1981). According to Chang and Page (1985), trace organics are found in natural waters and due to their low concentration probably do not significantly interfere with the utilization of the water. Nevertheless, the occurrence of many potentially toxic synthetic organic chemicals (such as halogenated hydrocarbons) that may reach the sources of drinking water are a matter of concern. Furthermore, some organic materials occur in water more frequently than other compounds. A list of trace organics found in secondary effluent, in different treatment plants, in the USA is presented in Table 2.9.

Table 2.9 Organic priority pollutants in secondary effluent (Overcash 1983)

Chemical group	Two compounds with highest concentration; average of 19 treatment plants (μg/l)		Range of concentration[a] reported (μg/l)	Annual organic chemical loading at 25 mm/week ($kg\,ha^{-1}\,yr^{-1}$)
Halogenated aliphatics	Methylene chloride	−48	0.16–48	0.63
	Methyl chloride	−12		—
Substituted aromatics	Toluene	−18	0.21–18	0.24
	Pentachlorophenol	−2.1		—
Phthalate acid esters	Bis(2-ethylhexyl) phthalate	−21	0.21–21	0.28
	Di-n-butil phthalate	−3.6		0.047
Polynuclear aromatics	Phenanthrene	−0.68	0.16–0.68	—
	Anthracene	−0.68		0.009

[a] For the chemical groups.

Table 2.9 presents data on priority pollutants, organic compounds of environmental concern and the yearly application at a traditional rate (Overcash 1983). Soil and plant aspects of trace organic application are discussed further in Chap. 3.

2.8.6 Suspended Solids and Clogging Hazard

Typical contents of suspended solids in raw, primary and secondary effluents are presented in Tables 2.1, 2.4 and 2.7, respectively. The range of values is quite wide and is dependent on the time allowed for sedimentation during the secondary treatment. The quantity of suspended solids may reach 0.1% (see Sect. 2.2). Secondary effluent to be used for drip irrigation requires filtering to reduce clogging hazard.

Low rate application irrigation systems with a large number of emitters of small orifice or narrow flow path, operating under low pressure, are very susceptible to clogging by suspended particulate matter chemicals and microbiological precipitates and bacterial slimes. Since clogging is greatly aggravated when more than one such contaminant is present in the irrigation water, there is no standard for acceptable levels of clogging contaminants. Nevertheless, Bucks et al. (1979) and Nakayama (1982; Table 2.10) proposed tentative water quality criteria for indicating emitter clogging hazards.

Abundance of both organic and inorganic particulate matter in the irrigation water may clog soil pores, particularly at the upper surface layer, thereby reducing infiltration rate and hydraulic conductivity. A 20–30% reduction in hydraulic conductivity may occur during the irrigation season when using unfiltered raw sewage effluent. Sandy soils are more susceptible to such clogging than clay soils.

Table 2.10 Clogging potential of irrigation water used in drip irrigation systems (After Nakayama 1982)

Type of hazard	Extent of hazard: Slight	Moderate	Severe
Physical			
Suspended soids (mg/l)	<50	50–100	>100
Chemical			
pH	<7.0	7.0–8.0	>8.0
Dissolved solids (mg/l)	<500	500–2000	>2000
Manganese (mg/l)	<0.1	0.1–1.5	1.5
Iron (mg/l)	<0.1	0.1–1.5	>1.5
Hydrogen sulfide (mg/l)	<0.5	0.5–2.0	>2.0
Biological			
Bacterial populations (maximum number/ml)	<10000	10000–50000	>50000

Table 2.11 Water-borne pathogens and their effect on health (Cowan and Johnson 1984)

Group	Genus	Effects on human health
Bacteria	*Salmonella*	Typhoid fever, paratyphoid fever, enteritis, salmonellosis, food poisoning
	Shigella	Dysentery
	Escherichia	Enteritis (pathogenic strains)
	Vibrio	Cholera, enteritis, food poisoning
	Clostridium	Gas gangrene, tetanus, botulism, food poisoning
	Leptospira	Leptospirosis
	Mycobacterium	Tuberculosis, skin granuloma
Viruses	Poliovirus	Fever, poliomyelitis, enteritis
	Coxsackievirus A	Headache, muscular pain
	Coxsackievirus B	Nausea, meningitis
	Echovirus	Diarrhoea, hepatitis
	Adenovirus	Fever, respiratory infection, enteritis, inflammation of the eyes (conjunctivitis), involvement of central nervous system
	Reovirus	Common cold, respiratory tract infections, diarrhoea, hepatitis
	Hepatitis A virus	Infectious hepatitis (fever, nausea, jaundice)
Protozoa	*Entamoeba*	Amoebic dysentery
	Giardia	Giardiasis
Helminths Trematodes	*Schistosoma*	Schistosomiasis (Bilharzia)
Cestodes	*Taenia*	Tapeworm infestation in man, in cattle eating eggs (*T. saginata* or *T. solium*, respectively), eggs develop into the Cysticerus stage
Nematodes	Ascaris	Roundworn infestation
	Anchylostomum	Hook worm infestation
	Heterodera	Potato cyst eelworm

Table 2.12 Expected removal of excreted microorganisms in various wastewater systems (WHO Scientific Group 1989)

Treatment process	Removal (log_{10} units) of			
	Bacteria	Helminths	Viruses	Cysts
Primary sedimentation				
Plain	0–1	0–2	0–1	0–1
Chemically assisted[a]	1–2	1–3[g]	0–1	0–1
Activated sludge[b]	0–2	0–2	0–1	0–1
Biofiltration[c]	0–2	0–2	0–1	0–1
Aerated lagoon[c]	1–2	1–3[g]	1–2	0–1
Oxidation ditch[b]	1–2	0–2	1–2	0–1
Disinfection[d]	2–6[g]	0–1	0–4	0–3
Waste stabilization ponds[e]	1–6[g]	1–3[g]	1–4	1–4
Effluent storage reservoirs[f]	1–6[g]	1–3[g]	1–4	1–4

[a] Further research is needed to confirm performance.
[b] Including secondary sedimentation.
[c] Including settling pond.
[d] Chlorination or ozonation.
[e] Performance depends on number of ponds in series and other environmental factors.
[f] Performance depends on retention time, which varies with demand.
[g] With good design and proper operation the recommended guidelines are achievable.

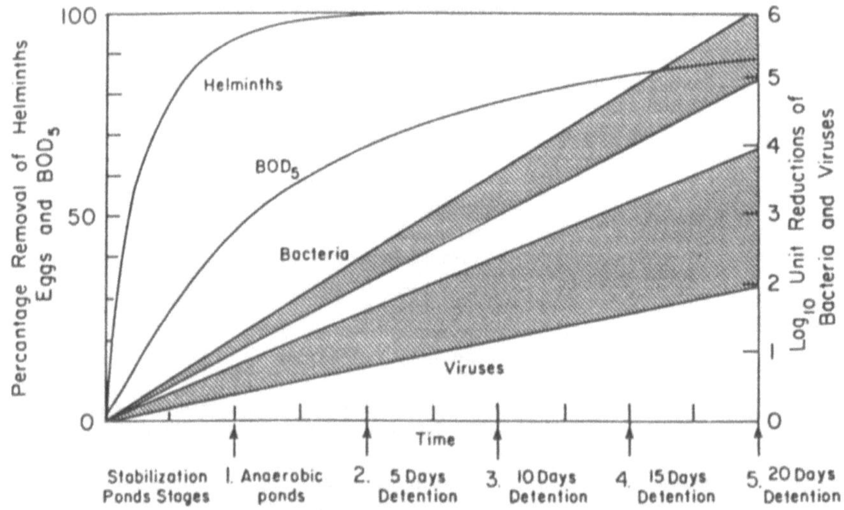

Fig. 2.4 Generalized removal curves for BOD, helminth eggs, excreted bacteria and viruses in waste stabilization ponds (Shuval et al. 1986)

2.8.7 Pathogens

Pathogenic organisms (bacteria, viruses, protozoa and parasitic worms) are found in raw sewage in various quantities depending on local conditions, e.g., population density and its sanitary habits and rate of disease and season of the year (Gerba 1983; Shuval et al. 1986; Pescod 1986; Shuval 1987). Most common pathogens associated with raw sewage are shown in Table 2.11.

Sewage treatment reduces the level of pathogens in sewage effluent to a great extent. The removal of pathogens by different treatment methods is shown in Table 2.12. In a recent publication Shuval et al. (1986) were able to show that stabilization pond treatment reduces the number of some pathogenic organisms even better than conventional biological treatment methods (Fig. 2.4).

The potential health hazard involved in the use of sewage effluent for irrigation of agricultural crops is the main effluent property used for its classification. The suitability of sewage effluent for agricultural irrigation is based on their microbiological quality. This effluent property indicates what crops can be grown in restricted irrigation, and if (in case of advanced treatment) the effluent is suitable for unrestricted irrigation.

Specific guidelines suggested for effluent irrigation and relevant soil, plant and environmental aspects are discussed in Sect. 3.9 below.

Chapter 3 Effect of Irrigation with Treated Sewage Effluent on Soil, Plant and Environment

3.1 Salinity and Sodicity

3.1.1 Determination of Soil Salinity

The total salt concentration (C) may be expressed in mol/l or mg/l, or in terms of the electrical conductivity (EC) of the water in dS/m or mmho/cm (1 mmho/cm = 1 dS/m). The average relationship of EC to concentration C in mg/l for many water sources in the EC range, 0.1–5.0 dS/m is:

$$EC = 0.00156\,C, \tag{3.1}$$

and $EC = 0.1\,C$ when C is mmol(charge)/l; [1 mmol(charge)/l = 1 milliequivalent/l].

It should be stressed that these equations are empirical and may not fit every single water source. The effect of the total concentration of soluble ions in the soil water on crops is referred to as the solute or osmotic effect. The average relationship of EC (in dS/m) to the osmotic pressure (OP, in MPa units) of the solution is given by

$$EC = 0.28\,OP, \quad \text{or} \quad OP = 3.6\,EC \tag{3.2}$$

For demonstrating yield response, soil salinity is usually expressed as the electrical conductivity of the soil saturated paste extract (EC_e). EC_e is used since the electrical conductivity of a solution is an easily, rapidly and reproducibly measured parameter, and because for many soils the soil saturated paste water content (θ_s) is a constant multiple of the water content at field capacity (θ_f), i.e., $\theta_s = 2\theta_f$.

A saturated paste of medium- to heavy-textured soils is obtained by mixing distilled water with a soil sample until the surface of the resulting paste glistens and the paste flows slightly when tipped. The paste is then placed in a funnel and the solution extracted by suction. Soluble ions may also be determined in the extract.

When many samples of a given soil material of known characteristics are to be analyzed, a 1:1 ratio of soil to water is easier to obtain and will result in a larger volume of extract. This method is not suitable for soil samples containing a slowly soluble salt such as gypsum.

The saturation extract method requires soil sampling and in vitro analysis, and is therefore slow and cumbersome. Some methods are now available for in situ determination of EC. They are:

1. Soil suction cups – solution samples are extracted with porous ceramic vacuum extractors in the field. Extraction must be done within a reasonable time following irrigation before root water uptake dries the soil below field capacity.
2. Salinity sensors – an electrode pair embedded in a porous ceramic disc is placed in contact with the soil and used to measure EC of the soil solution directly. This method is supposedly insensitive to changes in soil water content, inasmuch as the air entry value of the porous ceramic disc is high. However, measurements in dry soils are unreliable.
3. Four probe – soil resistivity can be measured using four probes, whereby an electric current is passed through the two outer probes while the two inner probes are used to measure the bulk soil resistance to current flow. The electrical conductivity of the soil solution can be estimated from the bulk soil resistance. The method is sensitive to changes in soil water content and therefore measurements should be taken at a constant soil water content (e.g., field capacity).
4. Electromagnetic induction – when an electromagnetic field is imposed on a soil, it induces a secondary electromagnetic field proportional to the EC of the soil, and thus can be used to measure it.
5. Time domain reflectometry (TDR) – the dielectric constant of many soils is uniquely related to the soil water content, and thus can be used to estimate it. Soil salinity attenuates the voltage pulse used to measure the dielectric constant in proportion to the salinity level. Thus, both water content and salinity may be measured simultaneously. This is a new and promising method, as yet untried on a large field scale (Dasberg and Dalton 1985).

The EC values obtained by all the above methods may be converted to EC_e using the water content of the saturated paste. A detailed description of these methods and their use is given by Rhoades and Oster (1986).

3.1.2 The Control of Soil Salinity

The increased salt concentration in the soil following irrigation is a direct result of evapotranspiration. All irrigation waters contain salts; however, whereas the water is taken up by plant roots and evaporates, most of the salts remain in the soil. If the water applied is just sufficient to satisfy the crop water requirement, salt will accumulate in the soil until water uptake ceases. Once the crop can no longer absorb water due to low osmotic potential of the soil solution (i.e., high salt concentration) accumulation will virtually cease. Salinity increase and reduced water uptake are interrelated, i.e., a higher salinity level will result in a lower uptake rate, whereas a higher uptake rate will result in faster salt accumulation. For a given irrigation water application (D_i), enhanced soil salinity will result in an increase in the fraction of the irrigation water moving beyond the root depth (drainage water; D_d). This fraction is called the leaching fraction, LF, defined as

$$LF = \frac{D_d}{D_i}. \tag{3.3}$$

Water movement through soil and the normal wetting and drying cycles result in dissolution and weathering of soil minerals, increasing the salt concentration in the soil solution to above the concentration in the irrigation water, C_i. During the wetting and drying cycles, the temporary increase in salt concentration in the soil solution may result in salt precipitation. Whether there is dissolution or precipitation will depend on C_i, LF and the salt composition. Excessive salt accumulation in the soil can only be prevented through leaching with surplus irrigation water or rainfall. Thus, leaching may be considered the key to the successful use of irrigation water containing high solute levels.

A satisfactory estimate of the leaching fraction may be obtained from a long-term study of the salt balance. Consider the condition when rainfall is low, salt precipitation in the soil is equal to the dissolution, salt uptake by crops is negligible, there is no addition of salt or water from outside the irrigated area, and steady-state conditions prevail. In this case, the amount of salt added with the irrigation water ($C_i D_i$) is equal to the quantity drained ($C_d \cdot D_d$) or

$$\frac{D_d}{D_i} = \frac{C_i}{C_d}. \tag{3.4}$$

Thus leaching fraction is equal to the ratio of the salt concentration in the irrigation water (C_i) to that in the drainage water (C_d) or, in terms of water consumptive use, D_e,

$$LF = 1 - \frac{D_e}{D_i}. \tag{3.5}$$

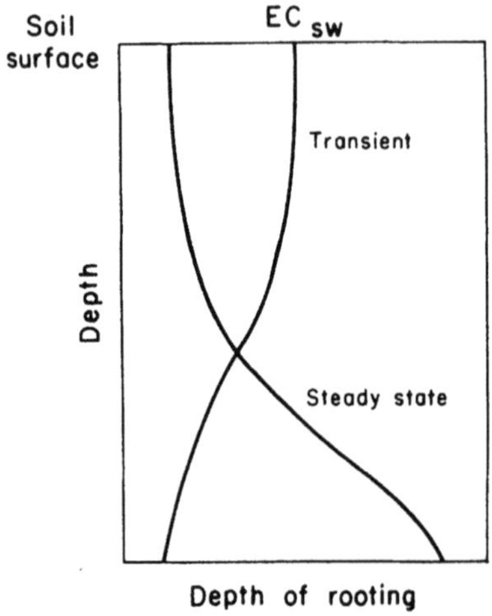

Fig. 3.1 Schematic representation of salt distribution in the root zone under transient and steady-state conditions, using the same quality irrigation water

Rainfall may be incorporated into Eq. (3.4) by averaging C_i over the total water application, $D_i + D_r$ where D_r is rainfall. However, its effect on leaching is indefinite since rainfall is not distributed uniformly during the year. When rainfall is seasonal, as is the case in a Mediterranean-type climate, the extent of leaching by rain will depend on soil physical characteristics (such as water storage and infiltration capacity). Coarse-textured soils, which have a low water holding capacity (WHC), will be more thoroughly leached than fine-textured soils, with a high WHC, since a given amount of rainfall will displace the pore volume in a given depth more times in the former soils.

Under steady-state conditions, soil properties have no effect on LF and, thus, on salt distribution in the soil. A typical salt distribution is shown in Fig. 3.1, with maximum concentration occurring at the lower part of the root zone. Since there is no water extraction below the depth of rooting, the soil solution cannot be concentrated any further. Thus, the maximum concentration that can be attained at the bottom of the root zone will be that concentration above which water uptake ceases.

The water extraction pattern has a dominant influence on the shape of the salt distribution profile. For deep-rooted crops, such as alfalfa, and low-frequency irrigation, salt will tend to accumulate deeper in the soil profile than for shallow-rooted crops, such as tall fescue, and high-frequency irrigation (Fig. 3.2). For seasonal crops under transient conditions, when initially the soil solution is nonsaline, the typical salt distribution will be as presented in Fig. 3.1. Under drip irrigation the typical pattern of accumulation will be as depicted in Fig. 3.3.

The above treatment of the control of soil salinity is based on the overall salt balance equation and, as such, is valid only for long-term average conditions. It treats the soil system only at its lower and upper boundaries, without

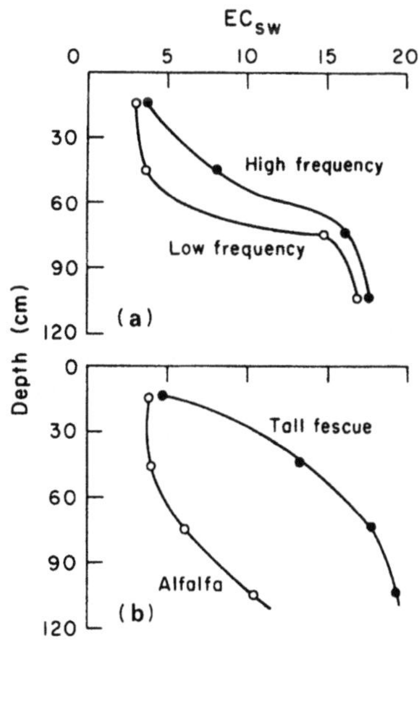

Fig. 3.2 a, b Salt distribution in the root zone under (a) high and low frequency irrigation (Bernstein and Francois 1973) and (b) for shallow-rooted (tall fescue) and deep-rooted (alfalfa) crops (Bower et al. 1969, 1970)

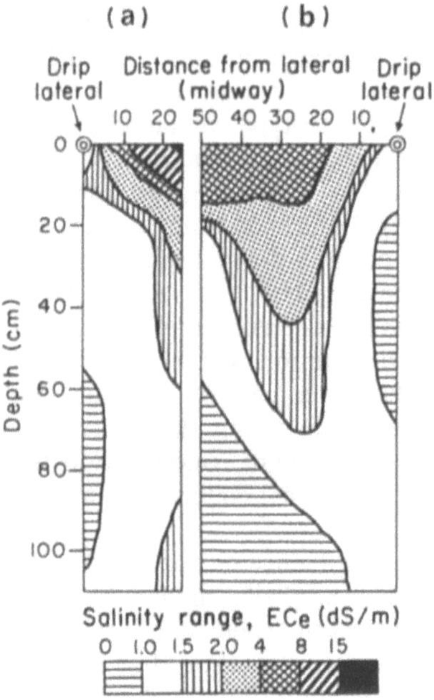

Fig. 3.3 a, b Two dimensional distribution of salinity in a loess soil under drip irrigation (a) when the distance between drippers is 50 cm and there is overlapping of wetting fronts midway between emitters and (b) when the distance between emitters is 100 cm without overlapping of wetting zones (Yaron et al. 1973)

considering the processes that take place within the boundaries. The approach was defined and publicized by the US Salinity Laboratory (USSL 1954).

A more rigorous and general treatment of the movement of salts and other hazardous chemicals in the soil is given in Sect. 3.2.

3.1.3 Soil Sodicity and Soil Physical Properties

High Na concentration may result in the deterioration of soil physical conditions through clay particle swelling and dispersion. This in turn causes a reduction in soil porosity, hydraulic permeability and infiltration rate, and destruction of the soil structure. When dispersion results in migration of clay particles to deeper soil horizons, the damage may be irreversible, which is the case for some sandy soils. The degree of swelling and dispersion depends on the nature of the predominant clay mineral: montmorillonite is the most sensitive, illite is moderately sensitive, and kaolinite is the least sensitive. Sodicity effects and particularly crusting and reduction in infiltration rate are more prominent in silt and silty loam soils.

The magnitude of the sodicity effect depends on the exchangeable sodium ratio (ESR) and on the total electrolyte concentration (EC) in the soil solution. The relation of the ESR to the sodium adsorption ratio (SAR) is given in the following equation:

$$ESR = K\,SAR \tag{3.6}$$

where

$$ESR = \frac{[Na]}{CEC - [Na]} = \frac{[Na]}{[Ca + Mg]}, \quad (3.6a)$$

Where the square brackets indicate concentration of exchangeable ions in the soil ($mmol\,g^{-1}$) and CEC is the cation exchange capacity of he soil, and where

$$SAR = Na/(Ca^{2+} + Mg^{2+})^{1/2} \quad (3.7)$$

with concentrations in solution expressed in $mol\,m^{-3}$. For most soils K is 0.012–0.015.

The interactive effects of the SAR of the irrigation water and its total salt concentration on soil permeability is illustrated in Fig. 3.4 (Oster et al. 1984). Irrigation water with a combination of SAR and EC_i values falling below the boundary line will cause impaired soil physical conditions. For example, the permeability of a soil irrigated with water having a total salt concentration as low as 3 mmol/l will be reduced to the same degree when the SAR of the water 15 $(mmol/l)^{1/2}$ as with water having SAR of 100 $(mmol/l)^{1/2}$ and salt concentration of 60 mmol/l. As a consequence, sodicity effects are uncommon in saline soils (McNeal and Coleman 1966), while the combinations falling above the line will have no or small detrimental effect.

When the saline soil solution becomes diluted following heavy leaching by winter rains, the soil becomes more sensitive to the sodicity effect of rapid particle dispersion and the consequent soil crusting. This process results in runoff and erosion during the winter and if the infiltration capacity is not restored, in difficulties in supplying sufficient water to the soil to satisfy crop needs during the next irrigation season (Shainberg and Oster 1978; Shainberg and Letey 1984).

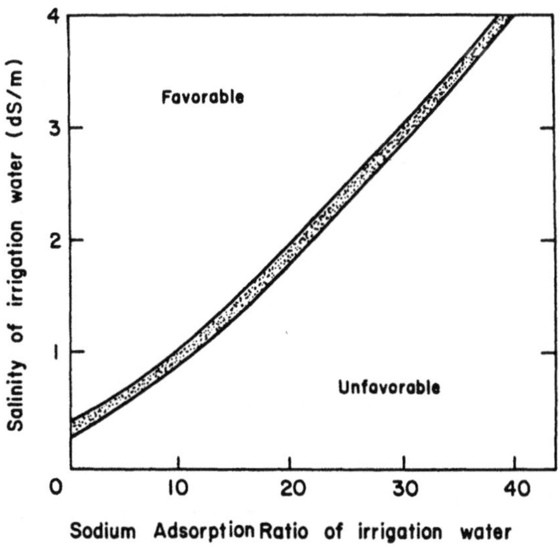

Fig. 3.4 Salinity and sodium adsorption ratio boundary indicating SAR and EC combination promoting favorable and unfavorable permeability conditions (After Oster et al. 1984)

Table 3.1 X-values calculated for various HCO_3/Ca ratios and ionic strength (After Suarez 1981)

	Ionic strength (I)[a]														
HCO_3/Ca	0.001	0.005	0.01	0.02	0.03	0.04	0.05	0.07	0.1	0.15	0.2	0.25	0.3	0.4	0.5
0.1	73.4	79.6	84.1	90.0	94.4	97.9	101.0	106.0	112.0	120.0	125.0	130.0	133.0	139.0	144.0
0.2	46.2	50.1	53.0	56.7	59.5	61.7	63.6	66.8	70.5	75.3	78.8	81.7	84.0	87.7	90.4
0.3	35.3	38.21	40.4	43.3	45.4	47.1	48.5	51.0	53.8	57.5	60.1	62.3	64.1	67.0	69.0
0.4	29.1	31.6	33.4	35.7	37.5	38.9	40.1	42.1	44.4	47.4	49.6	51.5	52.9	55.3	57.0
0.5	25.1	27.2	28.8	30.8	32.3	33.5	34.5	36.3	38.3	40.9	42.8	44.3	45.6	47.6	49.1
0.6	22.2	24.1	25.5	27.3	28.6	29.7	30.6	32.1	33.9	36.2	37.9	39.3	40.4	42.2	43.5
0.7	20.1	21.7	23.0	24.6	25.8	26.8	27.6	29.0	30.6	32.7	34.2	35.4	36.4	38.1	39.2
0.8	18.3	19.9	21.0	22.5	23.6	24.5	25.2	26.5	28.0	29.9	31.3	32.4	33.3	34.8	35.9
0.9	17.0	18.4	19.4	20.8	21.8	22.6	23.3	24.5	25.9	27.6	28.9	30.0	30.8	32.2	33.2
1.0	15.8	17.1	18.1	19.4	20.3	21.1	21.8	22.8	24.1	25.8	26.9	27.9	28.7	30.0	30.9
1.5	12.1	13.1	13.8	14.8	15.5	16.1	16.6	17.4	18.4	19.7	20.6	21.3	21.9	22.9	23.6
2.0	9.96	10.8	11.4	12.2	12.8	13.3	13.7	14.4	15.2	16.2	17.0	17.6	18.1	18.9	19.5
2.5	8.58	9.31	9.84	10.5	11.0	11.5	11.8	12.4	13.1	14.0	14.6	15.2	15.6	16.3	16.8
3.0	7.60	8.24	8.71	9.33	9.78	10.1	10.5	11.0	11.6	12.4	13.0	13.4	13.8	14.4	14.9
3.5	6.86	7.44	7.86	8.42	8.82	9.15	9.44	9.91	10.5	11.2	11.7	12.1	12.5	13.0	13.4
4.0	6.27	6.80	7.19	7.70	8.07	8.37	8.63	9.06	9.57	10.2	10.7	11.1	11.4	11.9	12.3
4.5	5.80	6.29	6.65	7.12	7.46	7.74	7.98	8.38	8.85	9.45	9.88	10.2	10.2	11.0	11.3
5.0	5.41	5.86	6.20	6.63	6.96	7.22	7.44	7.81	8.25	8.81	9.21	9.55	9.83	10.3	10.6
6.0	4.79	5.19	5.49	5.87	6.16	6.39	6.59	6.92	7.30	7.80	8.16	8.46	8.70	9.09	9.37
7.0	4.31	4.68	4.95	5.30	5.56	5.77	5.95	6.24	6.59	7.04	7.36	7.63	7.85	8.20	8.45
8.0	3.95	4.29	4.53	4.85	5.09	5.28	5.44	5.71	6.03	6.44	6.74	6.98	7.18	7.50	7.73
9.0	3.65	3.96	4.19	4.48	4.70	4.88	5.03	5.28	5.57	5.95	6.23	6.46	6.46	6.94	7.15
10.0	3.41	3.69	3.90	4.18	4.38	4.55	4.69	4.92	5.20	5.55	5.80	6.02	6.19	6.46	6.66

[a] $I = 0.0127\ EC_i$, where EC_i is the electrical conductivity of the irrigation water.

Effluent water may contain organic polyelectrolytes which bind divalent cations, increasing the SAR of the water and thus increasing the potential damage to soil hydraulic conductivity (Metzger et al. 1983).

Solution of the sodic soil problem or prevention of sodicity development is provided by the application of a source of Ca such as $CaSO_4$ (gypsum) or $CaCl_2$, or, in the case of calcareous irrigated soils, an acid or acid-forming materials such as H_2SO_4 (or S), which dissolves the soil lime present in large quantities in these soils. The amount of Ca added to the soil or irrigation water should be sufficient to reduce the SAR to safe levels. In most sewage effluents an SAR pickup of about 2–3 $(mmol/l)^{1/2}$ is partially compensated by an EC pickup, and if the supply water is acceptable for irrigation, no special problems are anticipated with the reclaimed water.

The effect of Na is aggravated by the presence of carbonates and bicarbonates in the irrigation water and by the concentration of CO_2 in the soil gas fraction. Carbonate and bicarbonate tend to precipitate with Ca, especially during the wetting and drying cycles of an irrigation season, thereby increasing the SAR (CO_3 is present only in water of $pH > 9$). Thus the SAR of water containing a high concentration of HCO_3, which is common for municipal effluent, must be adjusted for Ca precipitation. The most critical effect of high SAR values is on the permeability of the top soil (infiltration rate). The SAR for near-surface soil, where leaching is very high (LF = 1.0), can be calculated from the equation (Suarez 1981).

$$SAR_{adj} = \frac{Na}{Mg + Ca_{eq}} \tag{3.8}$$

where $Ca_{eq} = X(P_{CO_2})^{1/3}$.

Values of X are given in Table 3.1 for various values of HCO_3^-/Ca^{2+} in the irrigation water and of ionic strength. Values of ionic strength, I, can be calculated from the empirical equation $I = 0.0127$ EC mol/l (Griffin and Jurinak 1973) where EC of the water is given in dS/m. For surface soil layers P_{CO_2} may be taken as 7×10^{-4} or $(P_{CO_2})^{1/3} = 8.9 \times 10^{-2}$. A modification of Table 3.1 is given by Ayers and Westcot (1985).

3.1.4 The Crop-Salinity Production Function

The numerous available data on the salt tolerance of crops were summarized by Maas and Hoffman (1977) and more recently by Maas (1986). The crop salt tolerance can be expressed in terms of a threshold $EC_e(t)$, below which the relative yield (Y_r in %) is not affected by salinity, and above which it decreases linearly with increasing EC_e with slope (s) characteristic to each crop. Mathematically

$$Y_r = 100 - s(EC_e - t). \tag{3.9}$$

Figure 3.5 describes graphically this equation for four classes of crops (Maas and Hoffman 1977). The slope and threshold value for each crop were plotted in two-dimensional diagrams for the four main crop groups: field crops, forage

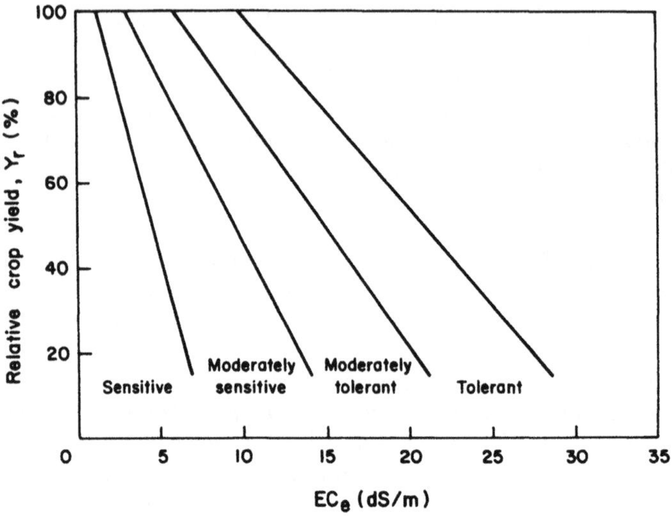

Fig. 3.5 Threshold and slope limits for salt tolerance rating groups (After Maas and Hoffman 1977)

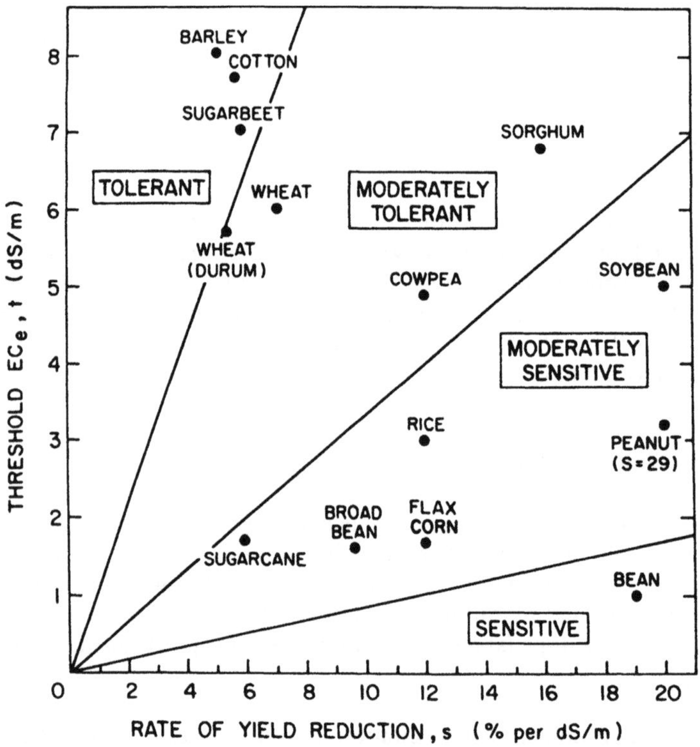

Fig. 3.6 Salt tolerance (threshold EC_e and % yield reduction with increasing salinity) of field crops (Maas 1986)

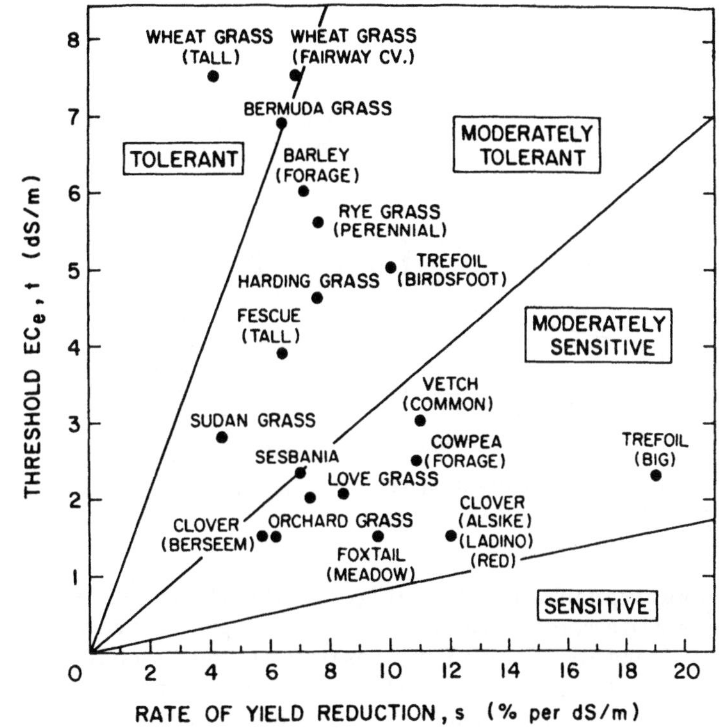

Fig. 3.7 Salt tolerance (threshold EC_e and % yield reduction with increasing salinity) of forage crops (Maas 1986)

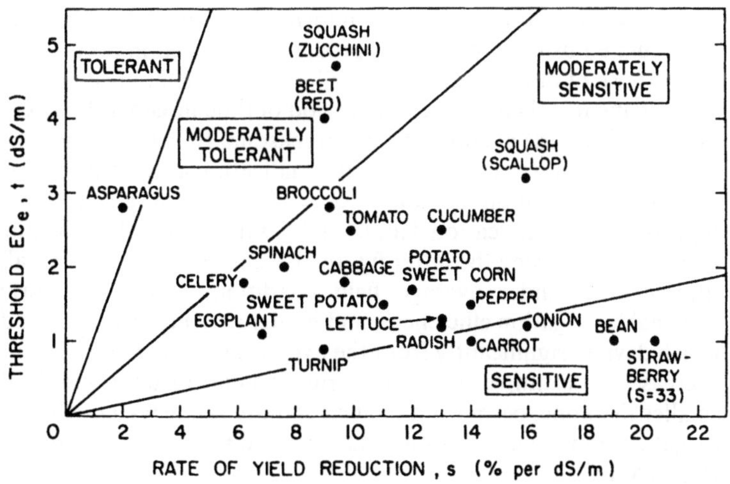

Fig. 3.8 Salt tolerance (threshold EC_e and % yield reduction with increasing salinity) of vegetables (Maas 1986)

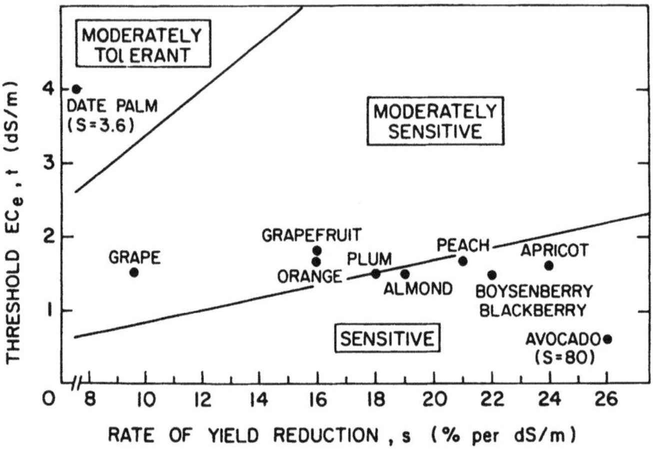

Fig. 3.9 Salt tolerance (threshold EC_e and % yield reduction with increasing salinity) of trees (Maas 1986)

crops, vegetables and trees (Figs. 3.6–3.9). The crops within each group are classified according to the four tolerance ratings suggested by Maas and Hoffman (1977) and are shown in Fig. 3.5 as typical salinity response curves, described by Eq. (3.9). Some crops for which precise threshold and slope information are not available are also listed within appropriate tolerance limits on the basis of educated estimates. Table 3.2 presents a list of salt tolerance ranges for various ornamentals in terms of threshold salinity values (Maas 1986).

The data as presented in Figs. 3.6–3.9 cannot be directly used before the relationship of EC_e to the electrical conductivity of the irrigation water (EC_i) is known, as normally only the latter information is available. The values of EC_e can be calculated using any of the approaches described in Sects. 3.1.2 and 3.2. For the purpose of our discussion we will use a very simplified empirical approach. Measurements taken over a long period of time in Israel (Boaz et al. 1976) have shown that, on the average, the mean seasonal EC_e within the rooting depth for citrus irrigation was equal to the EC_i, indicating a twofold accumulation of salt during the irrigation season ($EC_e = 1/2EC_{SW}$). Ayers and Westcot (1985) suggested a multiplication factor of 1.5 rather than 1, indicating a threefold accumulation. Using the latter figure, one can calculate the EC_i allowed for any specific crop under average field conditions (optimum cultural practices, reasonable irrigation efficiencies and moderate temperatures). Under conditions which deviate significantly from the average, the allowable EC_i may increase or decrease. For example, using drip irrigation may result in a multiplication factor smaller than one, permitting the use of more saline water, while using sprinkler irrigation on warm, windy days may cause direct leaf injury and thus reduce the tolerable EC_i.

The equation describing the leaching factor [Eq. (3.4)] may be used to calculate the leaching requirement (LR) of any specific crop, subject to the

Table 3.2 Salt tolerance of ornamental shrubs, trees and ground cover (After Maas 1986)

Common name	Botanical name	Common name	Botanical name
Very sensitive (maximum EC_e 1–2 dS/m)[a]			
Star jasmine	*Trachelospermum jasminoides*	Pyrenees cotoneaster	*Cotoneaster congestus*
Oregon grape	*Mahonia aquifolium*	Photinia	*Photinia x Fraseri*
Sensitive (maximum EC_e 2–4 dS/m)[a]			
Pineapple guava	*Feijoa sellowiana*	Chinese holly, cv. Burford	*Ilex cornuta*
Rose, cv. Grenoble	*Rosa sp.*	Glossy abelia	*Abelia x grandiflora*
Southern yew	*Podocarpus macrophyllus*	Tulip tree	*Liriodendron tulipifera*
Algerian ivy	*Hedera canariensis*	Japanese pittosporum	*Pittosporum tobira*
Heavenly bamboo	*Nandina domestica*	Chinese hibiscus	*Hibiscus rosa sinensis*
Laurustinus, cv. Robustrum	*Viburnum tinus*	Strawberry tree, cv. Compact	*Arbutus* unedo
Crape myrtle	*Lagerstroemia indica*		
Moderately sensitive (maximum EC_e 4–6)[a]			
Glossy privet	*Ligustrum lucidum*	Yellow sage	*Lantana camara*
Orchid tree	*Bauhinia purpurea*	Southern magnolia	*Magnolia grandiflora*
Japanese boxwood	*Buxus microphylla var. japonica*	Xylosma	*Xylosma congestum*
Japanese black pine	*Pinus thunbergiana*	Indian hawthorn	*Raphiolepis indica*
Dodonaea, cv. atropurpurea	*Dodonaea viscosa*	Oriental arborvitae	*Platycladus orientalis*
Thorny eleagnus	*Elaeagnus pungens*	Spreading juniper	*Juniper chinensis*
Pyracantha, cv. Graberi	*Pyracantha fortuneatna*	Cherry plum	*Prunus cerasifera*
Moderately tolerant (Maximum EC_e 6–8)[a]			
Weeping bottlebrush	*Callistemon viminalis*	Oleander	*Nerium oleander*
European fan palm	*Chamaerops humilis*	Blue dracaena	*Cordyline indivisa*
Spindle tree, cv. Grandiflora	*Euonymus japonica*	Rosemary	*Rosmarinus officinalis*
Aleppo pine	*Pinus halepensis*	Sweet gum	*Liquidambar styraciflua*

Table 3.2 (*Continued*)

Table 3.2 (*Continued*)

Common name	Botanical name	Common name	Botanical name
Tolerant (maximum > EC_e 8)[b]			
Brush cherry	*Syzygium paniculatum*	Ceniza	*Leucophyllum frutescens*
Natal plum	*Carissa grandiflora*	Evergreen pear	*Pyrus kawakamii*
Bougainvillea	*Bougainvillea spectabilis*	Italian stone pine	*Pinus pinea*
Very tolerant (maximum > EC_e 10)[b]			
White iceplant	*Delosperma alba*	Rosea iceplant	*Drosanthemum hispidum*
Purple iceplant	*Lampranthus productus*	Croceum iceplant	*Hymenocyclus croceus*

[a]Salinities exceeding the maximum permissible EC_e may cause leaf burn, loss of leaves and/or excessive stunting.
[b]Maximum permissible EC_e is unknown. No injury symptoms or growth reduction were apparent at 7 dS/m. The growth of all iceplant species was increased by soil salinity of 7 dS/m.

limitation of Eq. (3.9), by replacing the value of C_d by the soil salinity value representing the salt tolerance of the crop. LR may be defined as the minimum fraction of the total amount of applied water that must pass through the root zone in order to prevent reduction in crop yield. Despite the fact that Eq. (3.4) has been used for a long time, there is still some ambiguity regarding the selection of an appropriate value for C_d. Van Schilfgaarde et al. (1974) suggested selecting a value twice the electrical conductivity of the soil water at which no yield is obtained [by extrapolating the salt tolerance data given in Fig. 3.5 to zero yield, using Eq. (3.9)]. Rhoades (1974) denoted C_d by ($5t - EC_i$) where t is the threshold value [see Fig. 3.5 and Eq. (3.9)] and EC_i is the electrical conductivity of the irrigation water. The EC_i also replaces C_i in Eq. (3.4).

Salt tolerance data are usually given in terms of the mean EC_e in the root zone (Figs. 3.5–3.9). This assumes implicitly that the effective salinity is the mean total salt concentration at the relevant soil depth. This assumption is supported by some empirical results (Shalhevet and Bernstein 1967; Invalgson et al. 1976; Meiri 1984), whereas other experimental data indicate that the effective salinity is the mean salinity normalized on the basis of the water uptake distribution (Bernstein and Francois 1973).

3.1.5 Total Salinity (Osmotic) Effect

As shown in Figs. 3.5–3.9, crops vary widely in their sensitivity to salinity. The differences among crop plants are in the threshold salinity, i.e., the highest salinity not yet causing yield reduction, and in the rate of yield reduction with increasing salinity (slope) beyond the threshold [s and t in Eq. (3.9)]. Crops with similar rates of yield reduction may vary in tolerance according to differences in threshold (cf., Bermuda grass, tall fescue and orchard grass), while crops having similar threshold may vary with regard to slope (cf., eggplant, radish and bean). Variation in threshold more clearly characterizes tolerance than variation in slope. The salinity level at which no yield is obtained (the intersect on the salinity axis) also characterizes the salt tolerance of crops, as pointed out above (see Fig. 3.5).

In general, high total salt concentration suppresses growth rate, resulting in stunted plants. External symptoms of toxicity, such as necrotic lesions or tip burn are not usually evident, except for a dark blue-green appearance and thickening of the leaves. Leaf burn, resulting in complete crop loss, occurs only under extreme salinity conditions and is prevalent only when plants are sensitive to a high concentration of a specific ion (see Sect. 3.1.7).

Historically, the term "osmotic effect" was coined to denote the effect of the total salt concentration, since the reduction in water uptake by plants exposed to high salinity was thought to be the cause of yield reduction. In the early 1960s, it was found (Bernstein 1961) that plant roots adjust osmotically to changes in the osmotic potential of the root medium. This osmotic adjustment resulted in the maintenance of a potential difference for water uptake and raised doubts as to the validity of the assumption that yield reduction was due to reduced water uptake. The reason for the yield reduction due to high salinity is still unclear. It may be a result of a general disturbance in metabolic processes, such as the synthesis of essential metabolites, or of the effects on plant membranes or the energetics of osmotic adjustment. Different plant organs may respond differently to salinity. Certain crops, such as corn or tomato, exhibit a similar reduction for both economic fruit yield and vegetative growth, while for others, such as cotton, barley and wheat, the effect on vegetative growth is greater than on fruit yield. In the case of rice, grain yield reduction is much more severe than the suppression of vegetative growth (Bernstein 1974).

Stage of Growth and Varietal Differences. Crops may be differently sensitive to salt effects at different growth stages. Many crops are especially sensitive during early seeding, immediately after emergence, and salinity at this stage may cause irreversible damage. For other crops, the duration of salinity is dominant over any specific sensitivity, i.e., the longer the plants are exposed, the more severe the damage, regardless of growth stage (Shalhevet 1970).

In the past, because of poor emergence in saline fields, germination was thought to be a sensitive growth stage. It has since been shown that many crops are not more sensitive during germination than during later stages, but are sometimes even more tolerant (e.g., rice, corn, peanuts, and lettuce). Sugar beet, however, is notably more sensitive during germination than during later stages.

Some crops (in particular barley, Bermuda grass, lovegrass, wild varieties of tomato, lettuce, soybean, rice and wheat) exhibit varietal differences in response to salinity, but for most cultivated crops such differences are exceptional.

Fruit Quality. Salinity may have some beneficial effects. For example, the suppression of vegetative growth in cotton may, under some conditions, promote cotton lint yield, and at very low concentrations, an increase in salinity may result in yield increase, as is the case for peanuts. The most important benefit, however, is the effect of salinity on fruit quality. The dry-matter content of tomato (Brix index) as well as that of sugar beet and carrot, has been shown to increase with increasing salinity, partially compensating for lower dry matter production. Temporary soil salinization has increased tomato dry-matter content without affecting yield. Thus, salinity may be used to enhance the quality of tomatoes, especially for processing.

3.1.6 Specific Ion Toxicity Effects

Some crop plants may be sensitive to high concentration of specific ions. The toxic ion causes yield reduction at concentration (osmotic pressure) level which has no effect on yield if a mixture of salts or another salt is present. The Cl, Na and

Table 3.3 Chloride tolerance limits of fruit crop cultivars and rootstocks (After Maas 1986)

Crop	Rootstock or cultivar	Maximum permissible[a] Cl in saturated extract mol/m^3
	Rootstocks	
Avocado	West Indian	8
(*Persea americana*)	Guatemalan	6
	Mexican	5
Citrus (*Citrus* spp.)	Sunki mandarin, grapefruit, Cleopatra mandarin, Rangpur lime	25
	Sampson tangelo, rough lemon, sour orange, Ponkan mandarin	15
	Citrumelo 4475, trifolate orange, Cuban shaddock, calamondin, sweet orange, Savage citrange, Rusk citrange, Troyer citrange	10
Grape	Salt Creek, 1613–3	40
(*Vitis* spp.)	Dog ridge	30
Stone fruit	Marianna	25
(*Prunus* spp.)	Lovell, Shalil	10
	Yunnan	8
	Cultivars	
Berries[b]	Boysenberry	10
(*Rubus* spp.)	Olallie blackberry	10
	Indian summer raspberry	5
Grape	Thompson seedless, perlette	20
(*Vitis* spp.)	Cardinal, black rose	10
Strawberry	Lassen	8
(*Fragaria* spp.)	Shasta	5

[a]Higher concentrations may cause leaf injury. For some crops these concentrations may also exceed the osmotic threshold.
[b]Data available for one variety of each species only.

B ions are most prevalent in effluent water which may be specifically toxic, in addition to some minor elements present in low concentration in some effluents.

Chloride. The most prevalent ion in irrigation waters is chloride. It causes specific damage in some crops, mainly in woody perennial fruit trees and ornamentals (as shown in Table 3.3). The values in the table need to be divided by 1.5 to convert them to Cl concentration in the irrigation water, and to be multiplied by the equivalent weight of Cl (35.5 mg/l) to obtain the concentration in mg/l. The common initial symptom is bronzing of the leaves, especially on the south side (in the northern hemisphere) of citrus trees. At a more advanced stage, the tip of the leaf dries and the drying spreads along the leaf margin until the leaf drops. The threshold leaf chloride concentration beyond which leaf injury appears is 3–5 g/kg in the dry matter. However, under certain conditions, symptoms do not appear even at concentrations as high as 10 g/kg.

Usually leaf injury is due to the direct wetting of leaves with saline irrigation water, and for this reason irrigation is applied under the canopy in orchards.

One of the causes of chloride injury is thought to be nutritional imbalance. Although there is considerable evidence that the increase in Cl^- uptake by some crop plants causes reduced NO_3^- uptake resulting in nitrogen deficiency, it has yet to be demonstrated quantitatively that the effect is significant and can be corrected by nitrogen fertilization. This nutritional imbalance is most likely to be of importance for very sensitive crops, such as avocado, which are injured by very low levels of chloride.

Sodium. The Na hazard is threefold: (1) sodium may be specifically toxic in the same way that Cl is; (2) it may cause nutritional imbalance through impaired Ca or K nutrition (Bernstein 1964), and (3) it may affect the physical conditions on the soil, resulting in poor aeration and inadequate water supply (see Sect. 3.1.3).

Specific toxicities are usually limited to fruit trees, where leaf injury occurs at equivalent concentrations similar to or lower than those given in Table 3.4. In Israel, Na in the irrigation water was found to be toxic to citrus at levels higher than 5 mmol/l, compared to threshold of 6 mmol/l for Cl. The first symptoms to appear are drying of margins of mature leaves and necrotic spots. The necrosis advances into the leaf between the veins until complete drying occurs and the leaf drops. The threshold Na concentration for leaf damage in fruit tree leaves is 2–5 g/kg in the dry matter.

The effect of Na on nutritional imbalance depends not so much on the absolute concentration of Na but on the ratio of Na to Ca plus Mg in the soil solution (SAR). This effect which occurs more in nonwoody crop plants (see Table 3.4; Pearson 1960) is related more to impaired Ca nutrition than to Na excess, is not so crop specific, and is more severe under low rather than under high salinity conditions (Lagerwerff and Holland 1960).

A high Na concentration can be tolerated if the soil solution contains at least 1 mol/m^3 Ca, which is normally the case in saline soils.

Sodium may effect the potassium nutrition of plants, since at high concentrations Na^+ may compete with K^+ uptake. There is no specific, unambiguous

Table 3.4 Tolerance of various crops to exchangeable sodium percentage (ESP) under nonsaline conditions (After Pearson 1960)

Tolerance to ESP and range at which affected	Crop	Growth response under field conditions
Extremely sensitive (ESP = 2–10)	Deciduous fruits Nuts Citrus, *citrus* spp. Avocado, *Persea americana*	Sodium toxicity symptoms even at low ESP values
Sensitive (ESP = 10–20)	Bean, *Phaseolus vulgaris*	Stunted growth at these ESP values even though soil physical conditions may be good
Moderately tolerant (ESP = 20–40)	Clover, *Trifolium* spp. Oats, *Avena sativa* Tall fescue, *Festuca elatior* Rice, *Oryza sativa* Dallis grass, *Paspalum dilatatum*	Stunted growth due both to nutritional factors and adverse soil conditions
Tolerant (ESP = 40–60)	Wheat, *Triticum aestivum* Cotton, *Gossypium hirsutum* Alfalfa, *Medicago sativa* Barley, *Hordeum vulgare* Tomato, *Lycopersicon* Beet, *Beta Vulgaris*	Stunted growth usually due to adverse soil physical conditions
Most Tolerant (ESP = more than 60)	Crested and Fairway wheat grass *Agropyron* spp. Tall wheat grass, *Agropyron elongatum* Rhodes grass, *Chloris gayana*	Stunted growth usual due to adverse soil physical conditions

evidence of this effect, owing to the strong preferential uptake of K^+ by most crop plants. Some crops, such as corn, potato and eggplant and some citrus rootstocks (sour orange) may exclude Na. In stone fruits, the exclusion may be temporary, since, as the trees mature and sapwood is converted to hardwood, Na^+ is released from the roots and lower stem and is transported up the stem (Maas 1986).

Boron. Boron is an essential nutrient for most crop plants but toxicity appears when its concentration in the irrigation water exceeds 0.5–1.0 mg/l. The range of optimum boron concentration between deficiency and toxicity is thus rather narrow.

Crops may be damaged by excessive amounts of boron without exhibiting any distinct symptoms, such as chlorotic or nectotic spots on the leaves. Normally leaves contain 30–100 mg B/kg. Pome and stone fruits do not accumulate high levels of B in their leaves, but develop characteristic twig dieback and gummosis.

The B tolerance of some important crops and ornamentals is presented in Table 3.5, according to the four relative tolerance ratings used for salinity effect (Maas 1986). In each group, a range of B concentrations is given since the sensitivity of crops to B varies with climatic conditions, soils and varieties. The values given indicate maximum permissible concentrations in the soil water

Table 3.5 Boron tolerance of agricultural crops (After Maas 1986)

Sensitive (0.5–1.0 mg/l)[a]	*Moderately sensitive* (1.0—2.0 mg/l)[a]
	Broccoli (*Brassica oleracea botrytis*)
Lemon (*Citrus limon*)[b,c]	Pepper, red (*Capsicum annuum*)
Blackberry (*Rubus* sp.)[b,c]	Pea (*Pisum sativa*)[c]
Avocado (*Persea americana*)[c]	Carrot (*Daucus carota*)
Grapefruit (*Citrus x paradisi*)[c]	Radish (*Raphanus sativus*)
Orange (*Citrus sinensis*)[c]	Potato (*Solanum tuberosum*)
Apricot (*Prnus armeniaca*)[c]	Cucumber (*Cucumis sativus*)
Peach (*Prunus persica*)[c]	
Cherry (*Prunus avium*)[c]	*Moderately tolerant* (2.0–4.0 mg/l)[a]
Plum (*Prunus domestica*)[c]	Lettuce (*Lactuca sativa*)[c]
Persimmon (*Diospyros kaki*)[c]	Cabbage (*Brassica oleracea capitata*)[c]
Fig, kadota (*Ficus carica*)[c]	Celery (*Apium graveolens*)[c]
Grape (*Vitis vinifera*)[c]	Turnip (*Brassica rapa*)
Walnut (*Juglans regia*)[c]	Bluegrass, Kentucky (*Poa pratensis*)[c]
Pecan (*Carya illinoinensis*)[c]	Barley (*Hordeum vulgare*)
Cowpea (*Vigna unguiculata*)[c]	Oats (*Avena sativa*)
Onion (*Allium cepa*)	Corn (*Zea mays*)
Garlic (*Allium sativum*)	Artichoke (*Cynara scolymus*)[c]
Sweet potato (*Ipomea Batatas*)	Tobacco (*Nicotiana tabacum*)[c]
Wheat (*Triticum aestivum*)	Mustard (*Brassica juncea*)[c]
Sunflower (*Helianthus annuus*)	Clover, sweet (*Melilotus indica*)[c]
Bean, mung (*Vigna radiata*)[c]	Squash (*Cucurbita pepo*)
Sesame (*Sesanum indicum*)[c]	Muskmelon (*Cucumis melo*)[c]
Lupine (*Lupinus hartwegii*)[c]	Cauliflower (*Brassica oleracea botrytis*)
Strawberry (*Fragaria* sp.)[c]	
Artichoke, Jerusalem (*Helianthus tuberosus*)[c]	*Tolerant* (4–6 mg/l)[a]
Bean, kidney (*Phaseolus vulgaria*)[c]	Tomato (*Lycopersicon esculentum*)
Bean, lima (*Phaseolus lunatus*)[c]	Alfalfa (*Medicago sativa*)[c]
Peanut (*Arachis hypogaea*)	Vetch, purple (*Vicia benghalensis*)[c]
	Parsley (*Petroselinum crispum*)[c]
	Beet, red (*Beta vulgaris*)
	Sugarbeet (*Beta vulgaris*)
	Very Tolerant (6.0–15.0 mg/l)[a]
	Sorghum (*Sorghum bicolor*)
	Cotton (*Gossypium hirsutum*)
	Asparagus (*Asparagus officinalis*)[c]

[a] Maximum permissible concentration in soil water without yield reduction (threshold concentration). Crops are listed in order of increasing tolerance.
[b] Sensitive to concentration < 0.5 mg/l.
[c] Tolerance based on reduction in vegetative growth.

that do not cause yield reduction (threshold concentrations). The values in Table 3.5 were obtained from sand culture studies, and are on the conservative side for normal field soils containing clay minerals.

Soils adsorb boron to various degrees, depending mainly on their clay content, soil water content and content of other soluble ions. Due to its low concentration and adsorption, it is extremely difficult to remove accumulated boron from soils; to do so requires large quantities of leaching water.

It has recently been shown (Keren and Bingham 1985; Keren et al. 1985) that the higher the clay content in the soil, the lower is the boron activity in the soil solution for any given amount of B added to the soil. Plant B uptake and the resulting injury is directly related to B activity in the solution. Thus, the higher the clay content, the lower will be the expected B injury for any given amount of B added with the effluent water.

3.1.7 Direct Injury to the Foliage

Salinity of the irrigation water may be specifically restrictive when the water is applied by sprinkling, especially under adverse weather conditions. Direct contact of water drops with the crop canopy and their subsequent evaporation causes salt deposition on the leaves and direct injury to the tissue (necrosis). The injury will be more severe if water is applied intermittently, as with impact sprinklers, than when applied continuously as with garden or minisprinklers, and if applied during the day rather than at night (Maas et al. 1982).

Leaf injury is very crop specific and depends more on leaf characteristics than on tolerance to salinity. Some crops, such as advocado and strawberry, may be sensitive to salinity yet tolerate high salt concentrations when sprayed on the leaves. Others, such as barley and sorghum, may be very salt tolerant yet easily absorb salts from foliar spray. Table 3.6 lists the rating for leaf injury

Table 3.6 Relative susceptibility of crops to foliar injury from saline sprinkling water[a] (After Maas 1986)

Sodium or chloride concentrations (mol/m^3) causing foliar injury[b]			
< 5	5–10	10–20	> 20
Almond	Grape	Alfalfa	Cauliflower
Apricot	Pepper	Barley	Cotton
Citrus	Potato	Corn	Sugar beet
Plum	Tomato	Cucumber	Sunflower
		Safflower	
		Sesame	
		Sorghum	

[a] Susceptibility based on direct accumulation of salts through the leaves.
[b] Foliar injury is influenced by cultural and environmental conditions. These data are presented only as general guidelines for daytime sprinkling.

during daytime irrigation as suggested by Maas (1986). Furthermore, since leaf injury is directly proportional to the frequency of water application by sprinkler irrigation (Maas et al. 1982) and less dependent on its duration, sprinkling should be as infrequent as practicable. One way of reducing leaf injury due to daytime sprinkling is by washing off the salt using nonsaline water at the tail end of the irrigation period. This approach has been shown to be successful for corn in Israel (Frenkel et al. 1984), but it requires the availability of a fresh water source in the field. Leaf necrosis is a direct result of the adsorption of Na and Cl ions into leaf tissue. Na and Cl are also absorbed by roots exposed to saline environment and may cause leaf necrosis in some crops.

3.2 Transport of Chemicals and Hazardous Materials in the Soil

3.2.1 Introduction

Groundwater pollution is a major concern under modern intensive agricultural conditions. Water flowing below the root zone of crops carries various contaminants to the vadose zone and eventually further to aquifers. An understanding of the transport of nutrients in the soil is important for a good irrigation-fertilization management, particularly when using sewage effluents.

Chemicals and other substances applied to the soil can be also displaced by runoff and soil erosion; they can be mixed in the soil by tillage operations or other mechanical processes. Although such processes may, in some cases, be of great significance we will consider here only the transport of chemicals in the soil solution. Extensive investigations during recent years have greatly increased our conceptual understanding of the major mechanisms affecting flow and transport processes. We know that migration of chemicals through the soil depends not only on conventional subsurface fluid flow but also upon a number of interactive processes by which their concentrations may be attenuated (e.g., adsorption, exchange, precipitation, complexation and chelations, microbial activity and plant uptake). Considered as a whole, the soil is a complex physical-chemical-biological system. Thus, any transport model, however complex in its mathematics, is a gross simplification of actual processes.

There are two basic approaches to modeling transport phenomena in soil: the classical mechanistic-deterministic one and the more recently developed stochastic approach. Fundamental to the deterministic model is the use of continuum-differentiable equations with measurable parameters, related to each other through basic principles of convservation of mass, energy and momentum. The system is completely and rigidly defined and a unique solution is obtained for each set of parameters. The presumption of the stochastic approach is that mass transport phenomena in field soils are intrinsically erratic processes which can be quantified by statistical methods. These approaches are appropriate for the description of the transport of salts as well as of fertilizers, organic compounds, bacteria and other suspended particulates.

Many models are available for analysis of flow and transport of various chemicals, using various approaches and a variety of methods to solve the model equations. However, except for limited problems, present models are not very effective in predicting solute behavior under actual field conditions, mostly because the physical data necessary to implement them cannot be obtained with a sufficient degree of accuracy, given the current state of the art. Nevertheless, in this section we will present a description of the basics of the mechanistic approach which gives an insight into the physical principles of solute transport. This may be useful in the comprehension of some matters dealt with in the following sections. We will only briefly mention some of the stochastic models.

3.2.2 Mechanistic-Deterministic Approach

We will consider only the isothermal transport of a nonvolatile solute as expressed by the convection-dispersion equation, often referred to as the miscible displacement theory (Wagenet 1983).

Solutes as well as suspended matter travel with the flowing soil liquid by what is generally termed convective flux or advection. Velocity distributions within and between pores of different sizes and shapes, and fluctuations of streamlines with respect to the mean flow direction, impose spreading or dispersion of the convectively transported constituents and formation of concentration gradients. This process, which is a consequence of advection, is often referred to as mechanical dispersion. Occurring simultaneously is the process of molecular diffusion. Diffusion is an active process, irrespective of advection, which produces a net flux of the solute toward a zone of lower concentration. The solute flux in the soil is the sum of these processes.

The volumetric liquid flux is generally described by Darcy's law, also assumed to be valid under unsaturated conditions:

$$q_l = -K(h)\,\mathrm{grad}\,H, \tag{3.10}$$

where q_l is the liquid (water) flux, $[LT^{-1}]$, defined as the volume of liquid passing through a unit area of soil in unit time, H is the hydraulic head, [L]. $H = h + z$, where h is the pressure head (positive in saturated soils and negative in unsaturated soils, where it is also, termed the matrix potential head) and z is the gravitational head. K(h) is the soil hydraulic conductivity, $[LT^{-1}]$ which has a constant maximum value in saturated soils and decreases with decrease of h (or the water content, θ) in unsaturated soils. The matrix potential head (h) and the volumetric soil water content (θ) are interdependent, the relationship commonly referred to as the soil water retention or characteristic curve. At field soil water contents lower than what is usually called "field capacity" the hydraulic conductivity becomes very small. Under such conditions molecular diffusion becomes significant. The total solute diffusional migration may be a few cm per year.

The one-dimensional downward flow equation, is obtained by combining Darcy's law, [Eq. (3.11)] with the equation of continuity (assuming a fixed solid matrix):

$$\frac{\partial\theta}{\partial h}\frac{\partial h}{\partial\theta}=\frac{\partial}{\partial z}\left[K(h)\frac{\partial h}{\partial z}\right]-\frac{\partial K(h)}{\partial z}+\phi_l, \tag{3.11}$$

where z is the vertical coordinate and ϕ_l is a source/sink term (e.g., plant water extraction).

Hence, predicting liquid flow in unsaturated soil usually requires knowledge of K(h) or K(θ) and (h). These are affected not only by the type of soil but also by the kind and concentrations of solutes present in the soil (e.g., SAR). Also, they are not easily measured in the field. Soils are in general heterogeneous, thus tests of hydraulic conductivity result in a scatter of data over several orders of magnitude, resulting in similar variations in the calculation of liquid flux. Some of the problems associated with the prediction of the liquid flow in unsaturated soils are presented by Nielsen et al. (1986).

The solute convective flux, j_c, $[ML^{-2}T^{-1}]$ is defined by:

$$j_c = qC, \tag{3.12}$$

where C is the concentration of the constituent in the soil solution, $[ML^{-3}]$.

Often in practice, with a liquid flux of only a few mm per day, the convective flux is dominant in solute transport (Bolt 1978). Hence, as a first approximation, the average distance a solute travels during a given time t can be estimated by

$$Z = qt/\theta. \tag{3.13}$$

Fick's first law, modified for a porous medium, defines the molecular diffusion flux, j_d, $[ML^{-2}T^{-1}]$:

$$j_d = -D_e\theta \operatorname{grad} C, \tag{3.14}$$

where D_e is the effective diffusion coefficient $[L^2T^{-1}]$, $D_e = D_o\gamma$, with D_o being the molecular diffusion coefficient in bulk water and γ a factor accounting for tortuosity and solute effects which depend on the degree of saturation. According to Nye and Tinker (1977), γ at saturation varies between 0.4 and 0.7. Kemper and van Schaik (1966) suggested that $D_e(\theta) = D_0 a e^{b\theta}$ (with $5.10^{-3} < a < 10^{-2}$ and $b = 10$).

Mechanical dispersion is usually characterized by the term dispersivity, λ, sometimes referred to as the characteristic mixing length of the soil. The value of longitudinal dispersivity typically ranges from about 10^{-3}–10^{-2} m for laboratory column experiments and from 10^{-2}–2 m for small field scale experiments under natural gradients (Jury and Sposito 1985). Available data on dispersivity determinations in large-scale field experiments indicate large variability.

Although mechanical dispersion is strictly a physical process associated with advection, it is commonly described in the same general form as Fick's law:

$$j_m = -D_m\theta \operatorname{grad} C, \tag{3.15}$$

where D_m is the mechanical dispersion coefficient, $[L^2T^{-1}]$ assumed to be proportional to the average pore velocity ($D_m = \lambda v$), ($v = q/\theta$).

Conventionally, the effects of both molecule diffusion and mechanical dispersion are combined into one term, the coefficient of hydrodynamic dispersion, $D_h = D_e + D_m$. Sometimes a third coefficient associated with a diffusive solute transfer between mobile and stagnant pore domains is added to the hydrodynamic dispersion coefficient (Davidson et al. 1983). As a result of hydrodynamic dispersion, the peak concentration of solute tends to decrease and at the same time a migration front will tend to move faster than it would by advection alone.

The total solute flux, j_T, is described as

$$j_T = j_c + j_h = qc + D_h \operatorname{grad} C. \tag{3.16}$$

The continuity equation applied to solute transport in soil is given by:

$$\frac{dA_i}{dt} = -\operatorname{div} j_T + \phi_i, \tag{3.17}$$

where A_i is the mass of solute in a unit volume of soil [ML^{-3}]. If A_i is partitioned between the liquid and solid phases then $A_i = \rho S + \theta C$, where ρ is the soil bulk density (mass of soil per unit volume of soil) and S is the amount of solute adsorbed per unit mass of soil (if the solute does not interact with the soil then $S = 0$). ϕ_i are solute source/sink rate terms not specifically included in S.

Retardation Processes. A most common approach to modeling migration of an interactive solute is attributing retardation to reversible adsorption or exchange processes. These are generally described by an equilibrium isotherm (S(c)), and are quantified by the distribution coefficient K_d, which is the slope of the isotherm, if linear [or if S(c) is nonlinear by a distribution function K_f]. The distribution coefficient is essentially a measure of a particular soil's affinity for a given contaminant and is an indication of the contaminant's mobility in the soil. K_d and K_f are normally determined by laboratory batch or column adsorption tests, mainly because field tests are complicated by hydrodynamic dispersion, which makes data interpretation difficult.

In actuality the application of K_d is severely limited for many reasons and is generally limited to low concentrations of the contaminant and should be determined using a solution, which has a composition similar to the transporting liquid (or irrigation water).

Introducing the distribution coefficient K_d into Eq. (3.17) and assuming steady state liquid flow, the one-dimensional downward convection-dispersion equation is:

$$R\frac{\partial C}{\partial t} = D_h\frac{\partial^2 C}{\partial z^2} - v\frac{\partial C}{\partial z} + \phi_i,$$

where $R = 1 + \rho K_d/\theta$ – the retardation factor, is the ratio of the sorbing contaminant front velocity to the average liquid flow velocity ($v = q/\theta$). Small changes of R strongly affect the time of arrival of the contaminant to groundwater. ϕ_i usually accounts for irreversible sources and sinks (e.g., chemical dissolution and precipitation, microbial transformations, consumption and degradation, adsorption and exudation by roots) and may be time-dependent.

Solutions of Eq. (3.18) with $\phi_i = 0$ have been developed for various boundary conditions (van Genuchten and Alves 1982). Nielsen et al. (1986) discuss the simplifying assumptions and difficulties with the use of Eq. (3.18).

Retardation of solute movement may also occur due to precipitation reactions, mostly affecting heavy and transition metals, but also affecting phosphates, sulfates and sulfides (Harmsen and Bolt 1982). The transfer of mass from solution to solids by precipitation depends on the chemical composition of the solution and not the mass existing in or on the solids, as in adsorption. Hence, it cannot be based upon a functional relationship such as Kd. Generally, precipitation and dissolution are the result of a number of differing but interdependent mechanisms that occur in the soil simultaneously. These processes are dependent not only on the concentration of the solute of interest but also on the concentration and form of other species, on pH and on temperature.

Other chemical reactions, such as acid-base, oxidation-reduction reactions, complex formation and chelation occurring in the soil may result in beneficial or detrimental effects on contaminant mobility, mainly through changes in valency and solubility of various compounds. Also, the contaminant may be altered to a form which increases or decreases environmental concern. Often these reactions are the consequence of microbial activity. The quantification of biological processes is very problematic. Nevertheless, using various simplifications, microbially induced transformations were included in models describing nitrogen transport (reviews by Iskander 1981; Nielsen et al. 1982), organic phosphates (Castro and Rolston 1977) and pesticides and toxic organic compounds (Rao and Jessup 1983).

Breakthrough curves are plots, against time or against the cumulative quantity of the leachate, of the solute concentration in the outflowing solution from a soil column to which specific solutions are added. Breakthrough curves provide important information concerning different physicochemical properties of the soil-solution system that govern solute retention in soil. Commonly, the C/C_0 ratio, where C_0 and C are the concentrations of the relevant solute in the influent solution and the leachate, respectively, are plotted against pore volumes of leachate (the ratio of leachate volume to the volume of solution contained in the pores of the soil column under investigation).

Theoretically, breakthrough is achieved when $C/C_0 = 1$. If piston flow is the only mechanism involved in solute displacement, then a vertical line at one pore volume would represent solute breakthrough. However, because longitudinal dispersion is a common phenomenon, the typical breakthrough curve of nonreactive solutes (e.g. Cl) has a sigmoid shape and $C/C_0 = 1$ is obtained at > 1 pore volumes. At 1 pore volume $C/C_0 = 0.5$ in ideal non-reactive solutes, and this is conventionally designated as the breakthrough point.

Under similar inflow rates (q) the smaller the dispersivity of the soil (λ) the steeper are the breakthrough curves of the nonreactive solute. The spreading effect, due to hydrodynamic dispersion, around the average travel distance Z [Eq. (3.13)] of nonreactive solute or one of a linear adsorption isotherm [Eq. (3.16)] may be estimated as $0.4\sqrt{Z}$, at volumetric fluxes (q) in the range of 1–100 cm day^{-1} (Bolt 1978). A shift of the breakthrough point ($C/C_0 = 0.5$) beyond

1 pore volume indicates occurrence of adsorption or other reactions by which the solute is retained in the soil. Weakly retained solutes (e.g, Cl) with a $C/C_0 = 0.5$ around the 1 pore volume point are easily leached. Such solutes move through the soil at virtually the same speed as the displacing solutions and approximately 2 pore volumes of water are sufficient for their removal from a soil column. On the other hand, many more pore volumes of leachate are necessary to remove strongly retained solutes such as $H_2PO_4^-$ anions. The extent of the retardation of a few important anions in soil, as reflected by their breakthrough curves, is present and discussed in Sect. 3.4.4.

Evidently there are numerous processes which may affect transport of chemicals in soils. By itself, each process depends upon a complex set of conditions which control its ultimate effect on a particular compound. If interactions and interdependence of the various processes are considered and the dynamic nature of conditions existing in the field is also included, the complexity may appear staggering. Consequently, numerous transport models are currently available. They involve varying simplifying assumptions, include differing mixes of terms depending upon the application of the model, and differ in the techniques used for solving the mathematical equations.

A major limitation to the implementation of the available models is the almost universal lack of experimental procedures to accurately determine quantitative parameters which have any meaning in an actual field application. Unfortunately, the most elegant and efficient model is only as good as the data used to solve it. Thus, for most cases transport of chemicals in the soil can only be predicted qualitatively, and only for specific problems.

3.2.3 Stochastic Approach

Under natural conditions, migration of chemicals generally takes place in complex, heterogenous and variable systems. Hence, albeit the mechanistic-deterministic models provide a framework for understanding the movement and retention of chemicals in a natural soil profile, their utilization to predict penetration depth or solute peak location hinges on how well their parameters represent the real flow system. The statistical approach attempts to deal with the supposedly inherent system randomness, which is due to the heterogeneity of the flow medium or uncertainties about the parameters with which the system is defined. Uncertainties may be due to insufficient or unreliable input data or the generally inadequate understanding of basic physical and chemical processes that are being described. A comprehensive account of the conceptual framework of the stochastic approach to groundwater flow and transport is presented by Dagan (1986).

In his review of available stochastic models for solute migration, Jury (1985) distinguished between four approaches of stochastic modeling, i.e., scaling theory, Monte Carlo simulations, stochastic continuum theory and stochastic-convective models. In all of them, however different in their assumptions and techniques, certain parameters central to transport are treated as random variables. Dagan and Bresler (1979, 1983), and Bresler and Dagan (1979, 1981) applied scaling theory (Warrick et al. 1977) to define the hydraulic transport parameters of the convection-dispersion equation to account for spatial variability of the field. Amoozegar-Fard et al. (1982) described Monte Carlo simulations, in which all of the parameters of the convection-dispersion equations are assumed to be random variables, and the equations are run with a large number of parameter values to obtain the mean and variance of the variables of interest. Jury (1982) suggested a stochastic-convective model, using a probability density function for the travel time of a nonreactive solute moving in a field soil under conditions of steady convective flow, with a single random variable being the net applied liquid at the surface. The transfer function model (Jury 1982) was extended by Jury et al. (1986) to include possible physical and chemical processes.

The newly developed stochastic models show promise, yet there are still problems with their application. For example, a major problem with scaling theory is the disagreement between different methods of estimating the statistical moments of the distribution of scaling factors, which are the basis for obtaining the variability of all other parameters of the models. Also, it seems that in cropped systems the impact of variability on the hydraulic properties is diminished. Monte Carlo simulations generally require much computer time and have been used so far on restricted cases of chemical transport, primarily to illustrate the implication of spatial variability on sampling requirements and calculation of mean values. The same is true for the transfer function model (Jury et al. 1986). It is not yet known how well it can describe concentrations in vertically heterogenous soils and whether it can satisfactorily estimate the flux. However, models of this type, which require only the minimum of input data, could be incorporated in stochastically-based management schemes.

3.3 Nitrogen

3.3.1 Introduction

Nitrogen is an essential nutrient element for living organisms, it is usually added in large quantities to agricultural lands to obtain high crop yields. Irrigation with secondary sewage effluent results in the addition of considerable amounts of N to the soil, depending on the content of N in the effluent and the volume of water applied. Consequently, the actual quantity added to crops varies widely and is greatly affected by climatic conditions, soil and effluent properties, cropping and irrigation managements.

Since effluent N eventually enters the N cycle in soils, irrigation with sewage effluent affects nitrogen uptake by the crop, the level of available N in the soil, gaseous N losses and N leaching below the rooting depth. Irrigation management determines the distribution of N application over time, since N is added in every irrigation with sewage effluent. It is clear that appropriate fertilizer management in effluent-irrigated soils is essential to obtain the most benefit from the presence of N in effluents, and to prevent from the application of excessive or deficient amounts of N.

3.3.2 Nitrogen Transformations in Soil

Level and Composition of Soil N. The total N level, as well as its distribution in the soil profile and its chemical composition, varies among soils. In the top layer of mineral soils, it ranges from 0.2–4 g/kg (Black 1968) and consists mainly (90%) of organic forms (Stevenson 1982a, b), while most of the remainder appears as clay-bound nonexchangeable NH_4. Usually only a small fraction of soil N is present in forms available to plants, namely, exchangeable and water-soluble NH_4^+, and NO_3^-.

The soil N substances are components of a dynamic soil N cycle (see Fig. 3.10), which is an integral part of the N cycle in nature (Stevenson 1982a, 1985).

The quantities of N added to soils through effluent irrigation can be similar to or even exceed those applied by standard fertilization during similar periods of time (Feigin et al. 1978, 1979). For instance, the amounts of N added in effluent irrigation experiments on cotton (*Gossypium hirsutum* L.) in Israel (Bielorai et al.

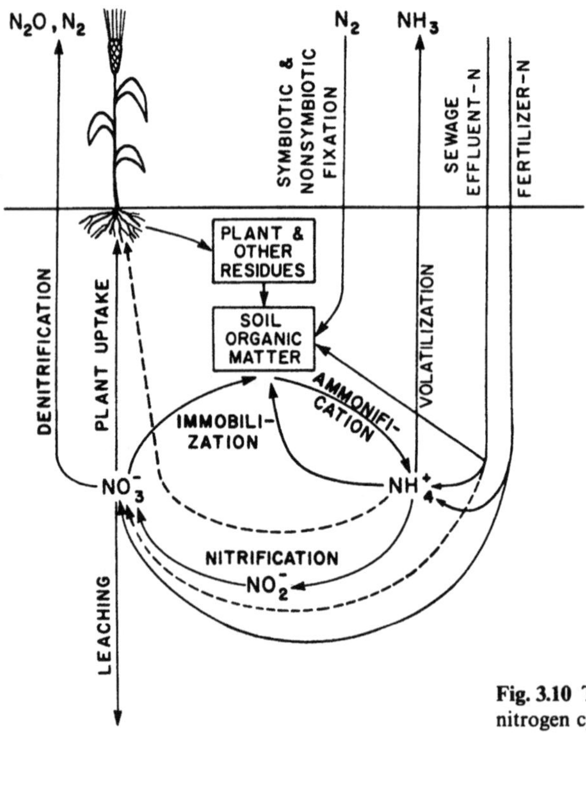

Fig. 3.10 The major components of the nitrogen cycle in the soil

1980, 1984; Dag et al. 1984; Feigin et al. 1978, 1984) were often greater than 200 kg ha^{-1} yr^{-1}, while the standard N-fertilization range for cotton in Israel is 120–180 kg N ha^{-1} yr^{-1}. The quantities of N given to Rhodes grass (*Chloris gayana* Kunth.) by irrigation with secondary sewage effluent were even greater, being about 500 kg N ha^{-1} yr^{-1}, which is similar to the amounts usually added to this perennial grass as fertilizer (Feigin et al. 1978). However, N fertilization is often necessary due to the presence of low N level in the effluent or when only small quantities of effluent are added, such as in supplementary irrigation of a crop (e.g., wheat grown under Israeli winter conditions).

As soon as the effluent N reaches the soil, it becomes part of the soil N cycle. The ammonium contained in the effluent, as well as that derived from organic N, is usually oxidized to NO_3^- by nitrification. Ammonia, derived from NH_4^+ under alkaline conditions, is susceptible to volatilization.

Both NH_4^+ and NO_3^- are taken up by crops, and eventually become immobilized as soil organic matter when crop residues are returned to the soil. The same process occurs in the tissue of soil microorganisms or their by-products.

Some of the NH_4^+ may be fixed into a nonexchangeable form by certain clays, e.g., the illite group (Nommik and Vahtras 1982). The significance of this process varies greatly among soils and its effect on the recovery of N added in sewage effluent is probably negligible.

Losses of NO_3^- by leaching below the rooting depth (eventually to the groundwater) and as gas (N_2 or N_2O) by denitrification are often large.

Ammonia Volatilization. The level of ammonia in effluent depends on the initial NH_4^+ content and the pH. When effluent containing NH_3 is used for irrigation volatilization may take place. In addition, NH_3 may be formed when effluent NH_4^+ reacts with alkaline soils. The relative concentrations of NH_3 and NH_4^+ in a solution depends on its pH, as indicated in the following equations:

$$NH_4^+ \rightleftharpoons NH_3\,(aq) + H^+, \tag{3.19}$$

$$\frac{[NH_3(aq)][H^+]}{[NH_4^+]} = K = 10^{-9.5}, \tag{3.20}$$

$$\log\frac{[NH_3(aq)]}{[NH_4^+]} = -9.5 + pH, \tag{3.21}$$

where NH_3(aq) is NH_3 in solution and [ ] denotes concentration. According to these equations, the concentrations of NH_3(aq) at pH5, 7 and 9 are 0.0036, 0.36 and 36%, respectively, of the total quantity of ammoniacal N in the soil solution. Loss of NH_3 from calcareous soils is considerably greater than from noncalcareous ones (Woldendorp 1968) and involves production of $(NH_4)_2CO_3$ (Fenn and Kissel 1973) or NH_4HCO_3 (Feagley and Hossner 1978). The fate of NH_4 added to a calcareous soil is represented by the overall reaction between $(NH_4)_2SO_4$ and $CaCO_3$.

$$CaCO_3 + (NH_4)_2SO_4 = 2NH_3 + CO_2 + H_2O + CaSO_4. \tag{3.22}$$

A quantitative model of NH_3 volatilization from calcareous soil was developed by Fleisher et al. (1987).

Mineralization-Immobilization. Secondary effluent contains organic N, sometimes at high levels (see Sect. 2.8). Organic N molecules are susceptible to microbial decomposition into simple inorganic N compounds, such as NH_4^+ and NO_3^-, which are available to plants (Fig. 3.10; Jansson and Persson 1982). Heterotrophic soil microorganisms, using organic compounds which contain N as a source of energy, are responsible for these mineralization processes.

The transformation of inorganic compounds (NH_4^+, NO_3^- and NO_2^-) into organic substances is defined as immobilization (Jansson and Persson 1982) and results from assimilation of inorganic N by soil microorganisms and plants. The mineralization-immobilization transformation is a dynamic part of the N cycle and has a great effect on the level of available N in the soil.

The rate of N mineralization or immobilization depends on the quantity and the quality of organic substances in the soil (including recently added material), on the inorganic N level, and on soil environmental conditions, e.g., moisture, aeration, temperature and pH. The concentrations of organic C and total N in the soil or crop residues, as well as the C/N ratio, are widely used to characterize these organic materials (Stevenson 1982b, 1985). A high level of available C will enhance microbial proliferation, and result in an increased demand for N, while decomposition of organic materials rich in N results in release of mineral N substances to the soil. Both processes are very dynamic and depend on the supply of C and N. The C/N ratio is therefore a very common parameter in the literature dealing with C and N transformation in soils.

The average C/N ratio in mineral soils is 10 (Stevenson 1982b, 1985), while the C/N ratio of biomass existing under well-balanced nutritional conditions in normal arable soils is about 25 (Jansson and Persson 1982). The incorporation into the soil of organic materials having C/N ratios greater than 25 (e.g., straw) results in N immobilization, while at lower C/N ratios (e.g., leguminous plant tissues) a release of available N takes place. Since total organic C, and total or organic N, relate to various substances which are very slowly decomposed, the E/N ratio (energy/nitrogen ratio), representing the readily decomposable materials, is often preferred over the C/N ratio (Jansson and Persson 1982).

Typical organic C/organic N ratios of sewage effluents are around 5 or lower. It may consequently be assumed that these materials are easily decomposable. Thus, the effect of irrigation with sewage effluent is similar to that of fertilization, since addition of a solution with a low C/N ratio will result in addition of available N to the soil.

Nitrification. Most of the effluent N consists of reduced N forms. The predominant N compound in sewage effluent is ammonium, which is a product of the breakdown of organic compounds during mineralization. However, since soil ammonium is susceptible to oxidation, NO_3^- is eventually formed and is the main available N form found in most soils after effluent irrigation.

The pathway of oxidation of NH_4^+ to NO_3^- is:

$$NH_4^+ + 1\tfrac{1}{2}O_2 \xrightarrow{6e} NO_2^- + H_2O + 2H^+ \quad (3.23)$$

$$NO_2^- + \tfrac{1}{2}O_2 \xrightarrow{2e} NO_3^- \quad (3.24)$$

The soil bacteria responsible for the first oxidation step ($NH_4^+ \rightarrow NO_2^-$) are mainly *Nitrosomonas* spp while *Nitrobacter* spp bacteria are involved in the second oxidation step ($NO_2^- \rightarrow NO_3^-$).

The effect of various factors, such as NH_4^+ and NO_3^- levels, aeration, moisture, temperature and pH on nitrification rate in soils was reviewed recently by Focht and Verstraete (1977) and Schmidt (1982).

The optimum temperature varies among soils, ranging from 20–40 °C (Schmidt 1982). The minimum and maximum temperatures for nitrification are 4 °C and 40 °C, respectively. Nitrification does not occur in dessicated soils, and the rate of the process greatly increases with increasing soil water content. Since good aeration is essential for uninterrupted nitrification, optimum conditions exist where both water and oxygen supply are adequate. The effect of pH on nitrification varies among soils. Whereas nitrification occurs within a wide range of pH, the optimum is pH 6–8.

Irrigation with sewage effluent affects the levels of NO_3^- in the soil, subsoil and groundwater, and the level varies with effluent quality and quantity, N fertilization, soil, crop and climatic conditions. As shown in Fig. 3.11, irrigation with sewage effluent greatly increased the level of NO_3-N in the top and deeper

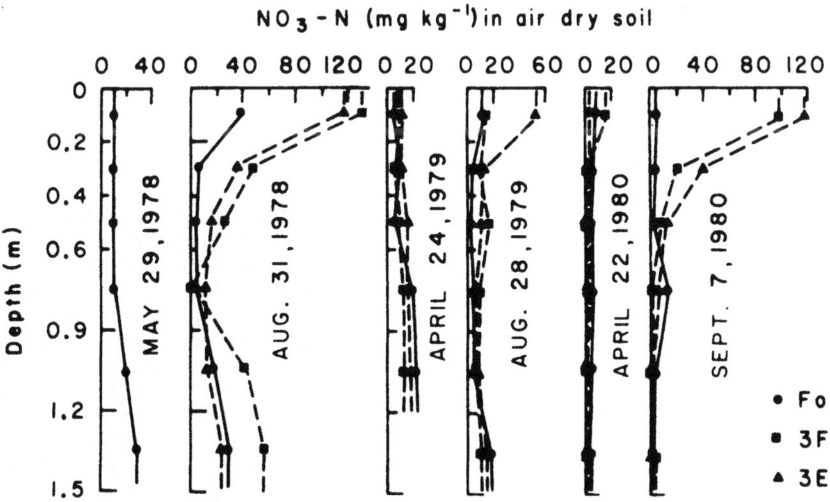

Fig. 3.11 Distribution NO_3 in soil profiles of the Zor'a drip irrigation experiment. ● fresh water without N addition; ▲ sewage effluent having an average of 50 mg N/1; ■ fresh water plus fertilizer to bring N, P, K level and application distribution to that of the effluent (Feigin et al. 1984)

layers of a clay soil planted with cotton (Feigin et al. 1984), compared to freshwater irrigated plots (in both cases drip irrigation was used and no fertilizer was added). The NO_3-N level fluctuations in the root zone were due to NO_3 uptake by plants, leaching into deeper layers (particularly during winter) and denitrification. The NO_3 increase was affected by enhanced mineralization of N in the soil in spring, and by irrigation with sewage effluent or by fertilization in summer.

Denitrification. The transformation of oxidized nitrogen into gaseous N compounds, i.e., denitrification, has recently attracted much attention, and its agricultural and environmental implications are now well recognized (Martin and Focht 1977; Payne 1981; Firestone 1982). The pathway of denitrification, $NO_3^- \rightarrow NO_2^- \rightarrow NO \rightarrow N_2O \rightarrow N_2$, whereby nitrate and nitrite are transformed into gaseous products (Fig. 3.10), is responsible for the loss of up to 70% (average 25–30%) of the nitrate found in soils under normal conditions (Hauck 1984a).

Denitrification takes place under anoxic conditions where available NO_3, NO_2, NO or N_2O exist in the soil together with microorganisms capable of

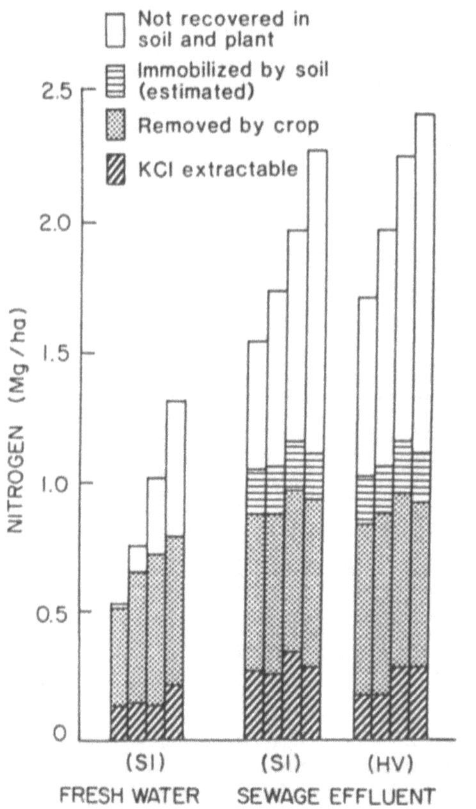

Fig. 3.12 Effect of sewage effluent and fresh water, both combined with four fertilizer levels, on nitrogen distribution in soil and cotton plants after 3 years of irrigation. *SI* Standard irrigation (3 irrigations per season); *HV* effluent quantity equal to 1.2 SI. Fertilizer N treatments: 0, 60, 120 and 180 kg ha^{-1} yr^{-1} correspond from *left to right* to the four columns within each of the three irrigation treatments. Each column represents the total N input which includes fertilizer N applied during the 3-year test, the amount of soil-available N (1.2 m depth) at the beginning of the test and the estimated potentially available N in soil (Feigin and Dag, unpubl. data)

reducing nitrogen oxides, and organic C substances are available as electron donors.

More than 20 genera of bacteria, including *Pseudomonas*, *Bacillus*, *Micrococcus* and *Achromonbacter* spp., have been identified as responsible for denitrification. There is a close relationship between available organic matter in the soil and denitrification (Bremner and Shaw 1958). Burford and Bremner (1975) found that mineralizable C or water-soluble organic C levels were satisfactory indexes for the capacity of a soil for denitrification; 1 μg of C was involved in the production of 1.17 μg NO_2 or 0.99 μg N as N_2. Addition of organic materials supplying available C to soil microorganisms often results in enhanced denitrification. Thus organic matter is added to effluent intended for groundwater recharge in order to reduce the level of NO_3 (Bouwer and Chaney 1974). Very high denitrification rates occur in soils treated with sewage sludge. Factors reducing gas exchange within soil pores and diffusion of O_2 into the soil tend to increase denitrification rates.

The optimum temperature for denitrification varies greatly among soils and ranges between 50 and 70 °C; the minimum temperature is 3–10 °C and the maximum about 75 °C. The optimum pH is 6–8; however, little denitrification has been reported at a pH as low as 3.5.

Table 3.7 Oxygen concentration in soil atmosphere (%) at four sampling dates (1975) in soil profiles under Rhodes grass (Feigin and Kipnis 1980)

Irrigations per week	1		2	
Nitrogen applied (kg/ha)	0	320	0	320
Depth (cm)				
		24 Aug		
20	20.2	19.0	19.5	10.1
50	14.8	19.4	8.6	(–)[a]
90	10.9	12.3	(–)	(–)
		29 Aug		
20	16.7	17.8	18.1	12.8
50	10.6	17.0	(–)	(–)
90	(–)	11.7	(–)	(–)
		31 Aug		
20	16.8	17.5	16.6	17.9
50	11.5	14.2	14.0	7.3
90	9.3	5.6	7.0	5.3
		4 Nov		
20	18.4	18.8	18.5	14.4
50	8.6	10.7	13.4	13.9
90	5.3	2.8	4.1	4.3

[a] Samples marked (–) were too wet to be analyzed.

Sewage effluent, particularly when only partially treated, contains considerable quantities of organic C (see Table 2.7); for instance, the level of C in effluent used to irrigate cotton in field studies reached 95 mg/l (Feigin et al. 1981). Consequently, enhanced denitrification rates may occur in effluent-irrigated soils.

A distinct increase in denitrification rate due to effluent irrigation was detected in a greenhouse experiment with a clay soil (Feigin et al. 1981), using tagged N to follow the recovery of N. Figure 3.12 presents results of N losses obtained in a field experiment in which irrigation with sewage effluent was compared with freshwater irrigation, each combined with four N fertilizer levels (0, 60, 120 and 180 kg N ha^{-1} yr^{-1}) (Dag et al. 1984). There was no loss of N from the nonfertilized freshwater irrigated plots, whereas there was 10–40% loss of added N from the fertilized plots. The corresponding range for the effluent-irrigated treatments was around 40–55%. Denitrification, NH_3 volatilization, and nitrate leaching probably accounted for most of the differences among the treatments.

The final N loss from effluent irrigated fields depends greatly on soil and environmental conditions. Losses of N from a Rhodes grass field at Zor'a (55% clay) were much greater (up to 50% of the N applied in effluent and fertilizer) than in a nearby cotton field (Feigin and Kipnis 1980). The quantity of effluent applied to the grass, and consequently organic C, was twice that given to the cotton; moreover, the irrigation frequency was much greater in grass (intervals of 3–7 days in grass and 2–3 weeks in cotton) and the grass field top soil remained wetter for longer periods. Furthermore, the grass was cut every 21 days during the active growing season, thus increasing available C in the soil through plant residues and dead roots. The O_2 level in the 0–0.9 m soil layer

Fig. 3.13 Effect of quality of irrigation water and fertilizer applied on NO_3-N distribution in several Zor's plots. —— fertilized lolium-clover field; --- effluent irrigated for 7 years and fertilized, cotton field; ------ unirrigated, fertilized wheat field; –·–·– effluent irrigated and fertilized Rhodes grass field (Feigin et al. 1978)

(Table 3.7) was often much lower in the Rhodes grass field than in the cotton field, where it was always greater than 14%. Tests carried out in a nearby commercial cotton field revealed high NO_3 levels in soil profiles (Fig. 3.13) to which either N-rich effluent or N fertilizer had been added. In contrast, the NO_3 level in Rhodes grass soil profile was always very low at the end of the growing season down to a depth of about 3 m (Feigin et al. 1978).

3.3.3 Nitrate Leaching from the Root Zone

High NO_3 levels in drinking water are hazardous, causing methemoglobinemia in infants. The permissible NO_3-N level is 10 mg N/l (45 mg NO_3/l; committee on Nitrate Accumulation in Soils 1972); the EEC guideline limit is 5.7 mg NO_3-N/l, with an acceptable level of 11.3 mg NO_3-N/l. An undesirable effect of NO_3 in surface water is enhanced proliferation of algae ("algae bloom"), which occurs in the presence of phosphorus and results in deterioration of potable water quality, even at very low N levels. Public awareness of environmental quality is reflected by the numerous articles on the impact of agrochemicals and other industrial chemicals on water pollution. Many of these papers deal with pollution of groundwater by NO_3 derived from sewage effluent.

Nitrates are susceptible to leaching from the plant root zone into deep soil layers and eventually into aquifers. Being an anion, NO_3 is not adsorbed by the negatively charged soil clay surfaces, and as its solubility in water is high, it is easily transported through the soil profile. In acid soils (pH < 5.5–6) or in the presence of amorphous minerals of volcanic origin, positively charged materials capable of adsorbing NO_3 are prevalent and NO_3 leaching declines.

The transport and distribution of nitrate in the soil profile in relation to various soil properties and crop management and its leaching to groundwater and surface waters has been the subject of many field studies. Letey et al. (1977) found a good correlation between the quantity of leached NO_3 and the product of drainage volume with fertilizer N applied.

Pratt et al. (1978) suggested an experimental method for estimating the amount of NO_3 leached from freely drained fields, by measuring the NO_3 and Cl concentrations in soil samples taken below the crop root zone and the Cl concentration in the incoming water. The method assumes Cl to be an inert ion, its only source being the irrigation water, and that below the root zone NO_3 and Cl behave similarly. Under steady state conditions the amount of NO_3 removed from the root zone in the percolating water, N_p, is

$$N_p = Cl_e \frac{(NO_3)_S}{Cl_S} \tag{3.25}$$

where Cl_e is the excess amount of Cl (total input of Cl in the irrigation water minus Cl removed by crop) and the $(NO_3)_S$ and Cl_S are concentrations in saturated soil extract below the root zone.

Adriano et al. (1972) calculated the transit time of the leached NO_3 in relation to the level of N fertilizer added and the level of irrigation applied using Eq. (3.26). Assuming NO_3 to be inert and neglecting hydrodynamic dispersion (see Sect. 3.2) the transit time (in years) until NO_3 reaches groundwater can be calculated as follows:

$$\text{Transit time} = \frac{Z\theta}{D} \tag{3.26}$$

where Z is the depth to groundwater, θ is the average volumetric soil water content and D is the annual volume of drainage water, in cm/yr.

Leachate volume D is usually of the order of magnitude of 0.1–1.0 m yr^{-1} and θ varies from a few percent in sandy soils to about 35% in clay soils (Pratt et al. 1978).

An empirical equation was used by McNeal and Pratt (1978) for predicting NO_3 leaching losses:

$$N_p = 0.2(N_iD)^{0.712}, \tag{3.27}$$

where N_p is the amount of leached N and Ni is N input, both in kg ha^{-1} yr^{-1} and D is volume of drainage water in (surface cm) ha^{-1} yr^{-1}.

Rible et al. (1979) estimated NO_3-N leaching below crop root zone in over 80 sites in California using the following equation:

$$N_p = \frac{(NO_3\text{-}N)D}{10}, \tag{3.28}$$

where N_p is the amount of N drained below the root zone (kg ha^{-1} yr^{-1}), NO_3-N in mg/l is the concentration in the soil solution below the root zone and D is the volume of drainage (cm/yr).

Various problems in estimating nitrate leaching from soils are discussed by Pratt et al. (1978).

A great deal of research on NO_3 leaching in sewage-effluent treated soils was carried out to develop efficient groundwater recharging technologies (Bouwer and Chaney 1974; Smith and Peterson 1982). High-rate effluent application (infiltration-percolation) involves the addition of large quantities of water, and unless the N level in the effluent used is negligible, large amounts of available N (much greater than those taken up by agricultural crops) are also applied. Hence, the allowable annual quantity of effluent conforming with the permissible NO_3-N level in leachate to groundwater must be considered [see Eq. (6.3), Sect. 6.1.2)]. Bouwer and Chaney (1974) reviewed methods using the addition of carbonaceous materials under anoxic conditions to enhance denitrification and thus prevent NO_3 leaching from top soil layers to groundwater. Other methods, aimed at increasing N uptake efficiency by crops and reducing residual NO_3 in the soil, are successful under low-rate land treatment methods (normal crop irrigation; Keeney 1982a).

3.3.4 Uptake of N by Crops

Agricultural crops take up large quantities of N (Table 3.8), and only rarely can high yields be obtained without the addition of available N, supplied mainly as fertilizer, manure or compost.

Table 3.8 Nitrogen yield in the total crop and in the harvested portion in relation to total yield in some important crops

Crop	Yield	Nitrogen yield: In total crop	Nitrogen yield: In harvested portions	References
	Mg/ha	kg/ha	kg/ha	
Alfalfa	18.0	510	—	Olson and Kurtz (1982)
Barley	3.76–4.5	75–125	72–85	Tucker and Hauck (1978)
Cotton[a]	3.5–5.9	110–300	60–140	Feigin et al. (unpubl. data)
Corn for silage	6.0–29.0	—	111–361	Tucker and Hauck (1978)
Corn	7.2–10.0	165–280	110–361	Tucker and Hauck (1978)
Hay[b]	4.5–18.0	94–438	—	Tucker and Hauck (1978)
Oats	3.0– 3.6	94–100	63–67	Tucker and Hauck (1978)
Pearl millet	—	500	—	Overman (1975)
Potato tuber	56.0	285	—	Olson and Kurtz (1982)
Rice (grain)	7.9	125	—	Olson and Kurtz (1982)
Rye	—	118–240	—	Goh and Haynes (1986)
Safflower	3.0	120	92	Tucker and Hauck (1978)
Sorghum	4.5–7.8	137–235	84–137	Tucker and Hauck (1978)
Soybeans	3.4	250	—	Broadbent and Reisenauer (1985)
Sugar beet	56.0–86.3	168–253	112–167	Tucker and Hauck (1978)
Sugar beet[c]	13.3–15.8	229–339	—	Feigin et al. (1979)
Sugarcane (stalks)	112.0	405	—	Olson and Kurtz (1982)
Wheat	2.7–7.7	72–262	56–172	Tucker and Hauck (1978)
Bermuda grass (coastal)	22.5	565	—	Olson and Kurtz (1982)
Kentucky bluegrass	5.5	—	147	Broadbent and Reisenauer (1985)
Reed canary grass	14	—	460	Wolcott and Cook (1976)
Rhodes grass	10.7–22.2	—	375–580	Feigin and Kipnis (1980)
Ryegrass	—	200	—	Overman and Ku (1976)
Rye	10.3–12.8	310–390	—	Overman and Ku (1976)
Hardwood forest		95		Wolcott and Cook (1976)

[a] Seed cotton.
[b] Nonlegume.
[c] Sugar yield.

For some crops (e.g., fodder) the harvested yield includes the whole plant, while in other cases (e.g., cotton) a large portion of the crop remains in the field and is reincorporated into the soil. The actual N uptake varies with variety and growth conditions, but general information on crop uptake, such as that included in Table 3.8, is useful for planning purposes.

The rate of N uptake ($kg\,ha^{-1}\,day^{-1}$) changes during the growing season and it is much greater during certain physiological stages of plant life. For example, the uptake of N by cotton plants, treated by different irrigation and fertilizer combinations, in a trickle irrigation experiment with sewage effluent (Feigin et al. 1984) was above 6 kg ha^{-1} day^{-1} in June, but below 3 kg ha^{-1} day^{-1} later in the season. Since the growth rate of annual crops declines greatly towards the end of the plant life cycle, addition of a large quantity of N by means of effluent irrigation too late in the season may result in an undesirable buildup of NO_3 in the soil. For this reason, among others, the timing of N addition is a very important factor in crop management. Hook (1983) concluded that actively growing vegetation reduces leaching, especially when the plant material is removed through harvest.

3.3.5 Effect of Overdose of N on Crop Production and Quality

"Response curves" describing the effect of increasing fertilizer levels on crop yield are useful tools for analyzing crop-fertilizer relationships (Cooke 1972; Haynes 1986). Various mathematical functions have been suggested for describing these curves for different experimental conditions. A typical response curve is shown in Fig. 3.14 (experiment B, well water). The initial N input increment results in increasing yield reaching a plateau where yield is constant. With very heavy applications (experiment A, effluent) yield is reduced, probably due to physiological disorders resulting from excessive NO_3 levels in plant tissues.

Figure 3.14 presents the effect of excesive N levels on seed cotton yields obtained under four different combinations of N fertilizer and sewage effluent irrigation. The response curves obtained following 6 consecutive years of effluent

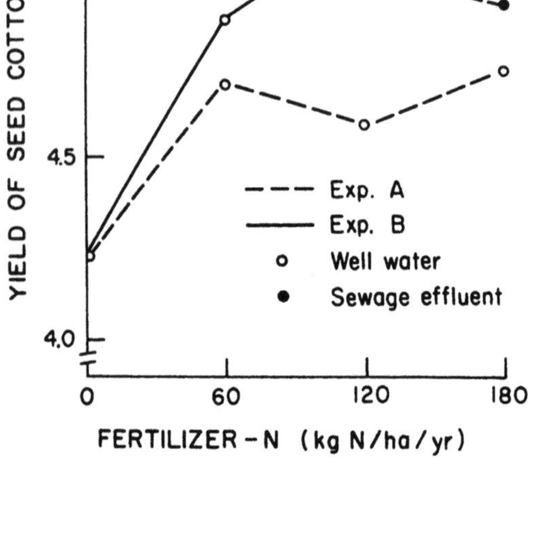

Fig. 3.14 Effect of quality of irrigation water and N fertilizer level on cotton yield (Dag et al. 1984)

irrigation (Dag et al. 1984) depict a small yield increase at the low N level (60 kg N ha^{-1} yr^{-1}) followed by a decrease at higher N fertilizer levels.

The nitrogen level in plant tissue affects carbohydrate metabolism (Black 1968). A high N content reduces the carbohydrate concentration in the plant and its storage organs. Feigin et al. (1978) noted a great reduction in the sucrose level in sugar beets due to the uptake of large quantities of N derived either from sewage effluent or fertilizer. Furthermore, high N levels in the beets cause processing difficulties. The total yield of sugar and the cane quality obtained from a sugarcane field irrigated by sewage effluent in Hawaii (Lau et al. 1975) were lower than those obtained using fresh water irrigation, even though the total yield was 11% greater.

Excessive N affects crop growth and yield. Enhanced vegetative growth as indicated by plant height (Table 3.9) and a reduction in yield was found in cotton plants to which excessive amounts of N were added through sewage effluent or freshwater amended with fertilizer. The tall plants had thinner shoots and severe lodging occurred, which is also a frequent problem in cereal crops which are heavily N-fertilized. Lodging was prevented in the 1979 and 1980 seasons, by the management changes (postponed first irrigation and somewhat reduced plant stand) discussed below (see Sect. 5.3).

Lint percentage is also reduced in cotton plants exposed to heavy N fertilization or to treatment with N-rich effluent (Bielorai et al. 1984; Dag et al. 1984). However, since in many cases the lint yield was not reduced, the effect of N was to increase seed yield, rather than to decrease lint yield.

Reduced cotton yield was obtained in a sewage-effluent irrigated field in California (Crites 1974) and was attributed to the excessive amounts of N (three times the normal value added as fertilizer). Many fertilization experiments (using well water) have shown that processing tomatoes were also greatly affected by

Table 3.9 Effect of sewage effluent, fresh water without fertilizer, and fresh water with fertilizer to meet effluent N, P and K levels, on cotton plant height (cm) (After Bielorai et al. 1984)

	Effluent				Fresh water		
	Without fertilizer				With fertilizer		
	Water (mm)						
Year[a]	350	440	515	440[b]	315	440	515
1978[a]	166b	176a	177a	141c	136c	163bc	168ab
1979	109bc	126b	136a	111c	101d	123b	130ab
1980	140bc	155a	156a	110d	145bc	154bc	145bc
Average	138bc	152ab	156a	121c	127bc	143ab	148ab

[a]Within years, figures followed by different letters differ by a 5% level of significance according to Duncan's Multiple Range Test.
[b]Fresh water without fertilizer.

excessive N. The enhanced vegetative growth resulted in longer fruiting and nonuniform ripening, which caused harvesting problems.

Succulence of plant tissue is also greatly influenced by N nutrition (Black 1968). In the presence of a high N supply, the formation of protoplasm and cells by carbohydrate utilization results in the development of large cells with thin walls. Internal breakdown of fruit during storage increases with the proportion of fruit with large cell volume. Black (1968) mentions cases of reduction in fiber crop quality (such as in hemp), due to tissue succulence caused by high N levels.

Delay in ripening of orange fruit have been associated with excessive N levels (Baier and Fraier 1973). Maynard et al. (1976) reviewed the undesirability of high levels of NO_3 in vegetable crops and concluded that the nitrate level in crop tissue increased as the rate of N fertilizer increased, and was higher with NO_3 addition than with NH_4 addition. The tissue NO_3 levels had not reached the range hazardous to adults.

3.4 Phosphorus

3.4.1 Introduction

Phosphorus is an essential nutrient element for all organisms and is added in large quantities to obtain enhanced crop yields. Secondary sewage effluent often has a high phosphorus content (Table 2.7 and Sect. 2.8), making it an important source of P for irrigated soils. Effluent can partially, or even completely, replace other P sources and prolonged effluent irrigation results in great savings in P fertilizer and has an important economic role.

The amount of P added to the soil through sewage effluent irrigation is usually not excessive. However, high loads of P are sometimes added, especially when the soil is used for disposal purposes. Excessive levels of available P may result in nutrient imbalances, such as Cu, Fe and Zn deficiencies (Ryden and Pratt 1980), necessitating soil or foliar application of micronutrients.

Application of P to the soil, as fertilizer, manure or in effluent, results in an immediate rise in the level of water-soluble P in the soil. However, due to adsorption and precipitation reactions taking place in the soil, the level of soluble P declines rapidly with time. Phosphorus uptake by plants also reduces the level of soluble P in the soil, but usually at a much slower rate. The actual level of water-soluble P in soil is usually very low, and the movement of P through the soil is restricted. Consequently, soil P compounds are usually considered immobile and P leaching and runoff have been considered negligible. Recent evidence, however, has established that the transport of even small quantities of fertilizer-derived P from agricultural soils into surface waters can induce eutrophication (algae bloom) in surface waters (Taylor and Kilmer 1980; Avnimelech 1984). Since 30 μg/l is sufficient to cause eutrophication in water, a yearly contribution of 0.3 g P/m^2 meets algae needs for inducing the process in a 10-m-deep lake. This quantity is equal to 1% of fertilizer P added to the soil at 3 g P m^{-2} yr^{-1} for a watershed area/lake area ratio of 10. Taylor and Kilmer (1980) concluded that excessive fertility levels in surface water, resulting from the transport of P through water runoff or soil erosion, is the only environmental problem associated with the pesence of P.

3.4.2 Phosphorus Concentration and Forms in Soils

The level of total P in soils varies greatly, lying between 0.2 and 0.8 g P/kg dry soil and 500–2000 kg P/ha in the top 0.2-m layer. The level P available to plant as determined by various chemical and biological methods (Olsen and Sommers 1982) comprises only a small fraction of the total P present. The concentration of water-soluble P is 0.03–3 mg/kg dry soil (Ryden and Pratt 1980), with most of the published values below 1 mg P/kg.

Typical levels of P in the subsoil solution of unfertilized soil range from 0.001 to 0.1 mg P/l (Hook 1983). The application of secondary effluents in which the concentration of P (mainly a orthophosphate) is 4–16 mg/l, may result in P penetration into subsoil layers. However, Hook (1983) reported only small changes in the P concentration in subsoils of many fields irrigated with secondary effluents at rates of 500–5000 mm/yr. The level of water-soluble P noted in various studies in subsoils of effluent irrigated cropland was 0.02–0.11 mg/l.

Soil P comprises organic and inorganic fractions, consisting of compounds characterized by different solubility and availability for various reactions, and with specific potential contributions to the pool of labile or available P in the soil. The percentage of each P fraction varies greatly. The organic fraction is 5–90% of the total P, with the higher percentages being typical for organic soils. The chemistry of soil P and the nature of its organic and inorganic fractions have been discussed widely (e.g., Dalal 1977; Parfitt 1978; Syer and Iskander 1981; Stevenson 1982a; Mengel 1985).

The principle inorganic P forms in soil are Ca orthophosphates, adsorbed orthophosphates and occluded phosphates (Lindsay 1979; Mengel 1985). The formation of Ca phosphates is promoted under alkaline conditions in the presence of a high Ca concentration. Since the pH and the level of Ca in calcareous soils are high, Ca phosphates are the predominant stable compounds in these soils. The different Ca-phosphate compounds found in soils are characterized by specific solubility and availability levels. The order of solubility of the common phosphate ions is: $H_2PO_4^- > HPO_4^{2-} > PO_4^{3-}$. The formation of stable phosphate compounds (tricalciumphosphate and apatite) is favored at high pH, while Al and Fe phosphates are formed at low pH.

The stability of inorganic P minerals in soils as a function of soil pH can be predicted using available information on their chemical properties, as shown by solubility diagrams proposed by Lindsay (1979). Figure 3.15 is a typical solubility diagram which relates the log $H_2PO_4^-$ or HPO_4^{2-} activity to soil pH, when Ca^{2+} content is $10^{-2.5}$ M. The figure consists of isotherms representing different P minerals. The area above each line indicates supersaturation that promotes formation of the relevant stable solids, whereas the area below

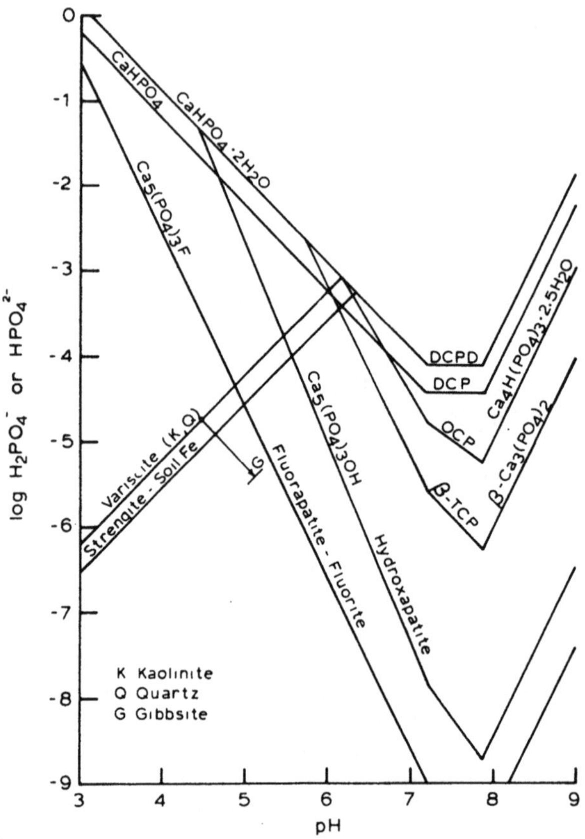

Fig. 3.15 Solubility of calcium phosphates compared to strengite and varsicite when Ca^{2+} is $10^{-2.5}$ M or is fixed by calcite and CO_2(g) at 0.0003 atm (Lindsay 1979)

indicates undersaturation promoting their dissolution. For instance, the formation of Fe- and Al-phosphates strengite [$(Fe(OH)_2H_2PO_4$] and variscite [$Al(OH)_2H_2PO_4$] is predicted at low pH, while the formation of Ca phosphates is favored at high pH. At pH 5–6, the solubilities Al-, Fe- and some of the Ca-phosphates are similar. Several phosphate minerals coexist within the pH range of 6–6.5. Solubility diagrams are based on thermodynamic properties of the relevant systems, but do not take into account kinetic aspects. They are helpful in predicting the behavior of different P species in the soil under various pH conditions, but sorption reactions often predominate and experimental data are often not well explained by solubility diagrams (Ryden and Pratt 1980).

3.4.3 Phosphorus Transformations in Soils

Substances containing phosphorus such as fertilizer, organic amendments and sewage effluent, when added to the soil, soon become an integral part of the P

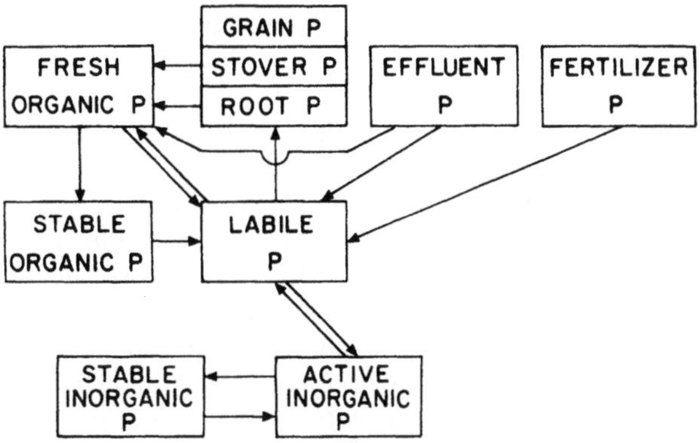

Fig. 3.16 Pools and flows of P in the soil-plant system (After Jones et al. 1984)

cycle in the soil. A simple model for the P cycle in the soil-plant system recently suggested by Jones et al. (1984) is used for describing the fate of P added by means of sewage effluent (Fig. 3.16). This model consists of pools representing the different P compounds, having specific chemical and physicochemical properties, and vectors indicating relationships between the pools.

Fertilizer P, consisting of soluble P, becomes part of the labile P in soils. Effluents, depending on the chemical properties of the P they contain, may also contribute to the pool of fresh organic P. Manure and other materials containing organic P may behave in a similar way. The labile P is available to plants and is therefore immobilized in plant material (grain, stover or root) and through microbial activity as fresh, and later also as stable, organic P. Some of the plant P, such as grain P, is removed from the field, while the rest of it is reincorporated into the soil, finding its way into the pool of fresh organic P. Decomposition of soil organic matter (fresh and stable pools) releases P back into the pool of labile P. The rate of this process depends on the nature of the soil organic matter, and on soil conditions such as pH, moisture, temperature and aeration. Only a small portion of the stable organic P is usually mineralized during the growing period of a typical annual crop such as corn, wheat or cotton.

The labile P peak concentration, which occurs immediately after the addition of soluble P to the soil, declines within a few hours or, at most, within a day or two, as P is being transformed into other forms, mainly by binding to the pools of active and stable inorganic P (Fig. 3.16). The relative proportions of the labile, active and stable inorganic-P forms are constant providing a constant supply of P as it is taken up by the crop.

3.4.4 Soil Processes Affecting P Movement

The predominant processes involved in the transformation of labile P to less soluble forms are sorption, precipitation and occlusion of inorganic P. The formation of insoluble Ca-phosphate compounds is the main process responsible for the reduction in the level of labile P in calcareous soils. The high concentration of Ca and the high pH, characteristic of these soils, promote the process.

Sorption processes are often the major explanation for the removal of labile P by noncalcareous soils mainly on Fe- and Al-oxides (Ryden and Pratt 1980). This is especially so in the case of sewage effluent in which the level of soluble P is usually much lower than that near fertilizer granules in the soil. Differences in the reaction rates of P in calcareous vs noncalcareous soils were found in North American soils (Barrow 1980). For most calcareous soils the P concentration is controlled by a solubility product.

The adsorption of phosphate on soil particles is usually represented by P adsorption isotherms. The isotherm shows an increase in the content of adsorbed P on the soil as the P concentration in the soil solution at equilibrium increases. The Langmuir equation [Eq. (3.29)] is widely used to describe phosphate adsorption on soil or soil constituents.

$$\frac{x}{m}=\frac{KCb}{1+KC}, \qquad (3.29)$$

where x is the weight of adsorbate (e.g., P), m is the weight of adsorbent (e.g., soil, clay or Al- and Fe- oxides), K is a constant related to the affinity of adsorbent, C is the equilibrium solution concentration of the adsorbate, and b is the maximum quantity of adsorbate that can be adsorbed. A typical Langmuir adsorption isotherm is shown in Fig. 3.17 (Bar-Yosef and Kafkafi 1978) in which the adsorption of a P on kaolinite clay gradually approaches a maximum. The Langmuir equation has been used extensively to measure P adsorption on soils, and diversified experimental results have been obtained for different soils. The greatest adsorption of P occurs in soils rich in Al- and Fe-oxides (sesquioxides), whereas in calcareous soils adsorption is less (Mengel 1985).

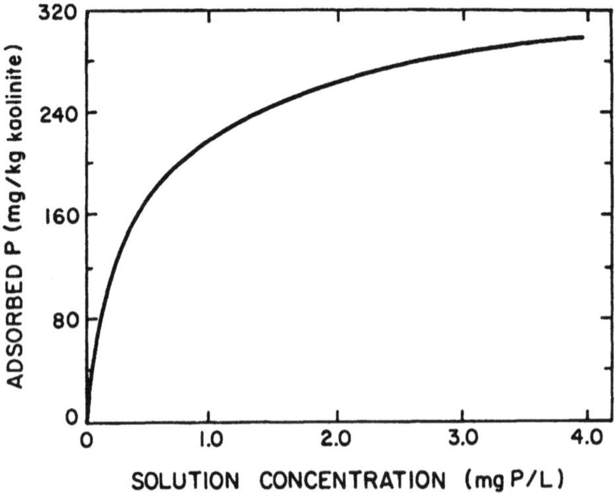

Fig. 3.17 Langmuir adsorption isotherm of P on kaolinite clay (Bar-Yosef and Kafkafi 1978)

The Freundlich equation [Eq. (3.30)] has also been used successfully for predicting P adsorption from solution (Bohn et al. 1985):

$$x/m = KC^{1/n}, \tag{3.30}$$

where K and n are empirical constants, x/m is the weight of adsorbate per unit weight adsorbent, and C is the equilibrium concentration of the solution adsorbate.

The difference in the retardation of $H_2PO_4^-$ in comparison with easily and moderately leached anions is shown below. Figure 3.18 presents breakthrough curves of anions most prevalent in soils (Bohn et al. 1985). Weakly retained anions (NO_3^- and Cl^-) are easily leached, and approximately 2 pore volumes of water are sufficient for their complete removal from the soil. On the other hand, 2 and 4 pore volumes, respectively, are needed to initiate leaching of the moderately retained anions SO_4 and H_3BO_3, and about 5 and 11 pore volumes, respectively, are needed for complete leaching. The $H_2PO_4^-$ anion is much more strongly retained and leaching only commences after 8 pore volumes have passed. After 16 pore volumes, only 60% of the anion is leached out. Experiments in which phosphate is added to soil by means of sewage effluent have shown that the breakthrough of P occurs only after 50–100 pore volumes have passed through the columns (see Sect. 3.4.5).

Different models used for predicting movement of P in soils (chemicals stability-, sorption-desorption-, kinetic-, and transport models) have been discussed by Enfield and Ellis (1983).

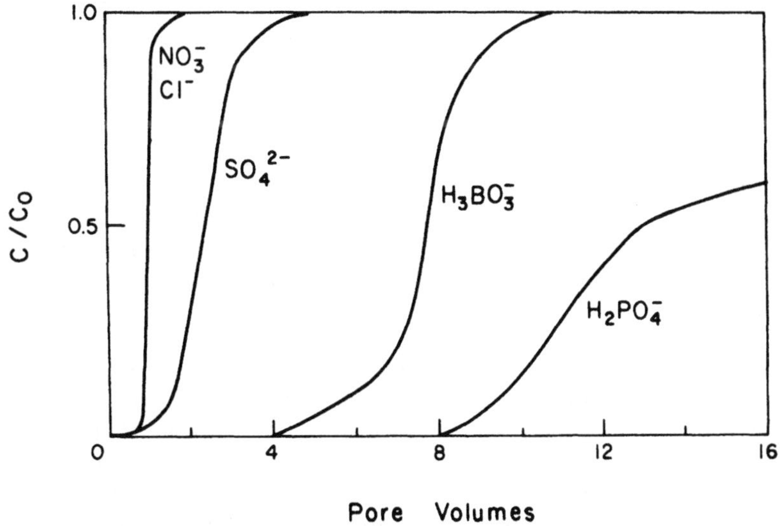

Fig. 3.18 Breakthrough curves of anions weakly, moderately and strongly retained by soils (Bohn et al. 1985)

3.4.5 Effect of Irrigation with Sewage Effluent on Soil P and P Movement in Soil Profile

P movement through effluent-treated soils has been studied under different conditions. Sawhney (1977) found that the breakthrough of P from a fine sandy loam and silt loam soil columns occurred after about 50 and 100 pore volumes of effluent, respectively, had passed through. Similar results were obtained by Nagpal (1985) in a column experiment using a 60-cm-deep gravelly silt loam soil. Equilibrium P concentration in the leachate was obtained after about 70 pore volumes of effluent had passed through the columns. Moreover, about 60–90% of the added P was retained by the soil even after 2.7 years of leaching, most of it in the surface soil. Phosphate transport through a calcareous loamy sand column (2.5 m long) was tested under extremely rapid infiltration (50–400 mm/day) of secondary sewage effluent (Lance 1977). The P concentration in the leachate was less than 1 mg/l during 200 days of 9-day flooding and 5-day drying cycles, after which it increased and finally leveled off at different concentrations depending upon the infiltration rate. Lance (1977) found that very large quantities of PO_4-P can be retained in calcareous sand. About 90% of the P added in the effluent (having a P level of 12 mg/l) was removed during initial stages of the test, decreasing later to about 75–80% of the added P. Adsorption of P continued even after the calculated initial sorption capacity of the soil was saturated, provided the infiltration rate was maintained below 150 mm/day.

The actual capacity of soil to adsorp P is often much greater than that predicted by laboratory tests. Alternate drying and wetting restore the adsorption capacity of soils (Ryden and Pratt 1980), and precipitation and transformation processes in soils also contribute to this discrepancy. Sawhney and Hill (1975) found that the movement of P, derived from a septic tank (point source), through a bed of sandy soil surrounding the tank was much slower than predicted. This was explained by the capability of sand to restore its P adsorption capacity. On the other hand, cases of faster-than-predicted movement of P were also recorded. Hill and Sawhney (1981) explained the more rapid transport of P through an undisturbed soil block treated with effluent by the existence of preferred pathways through which solutions moved faster than through homogeneous soils. Such pathways permit breakthrough of P to groundwater before the adsorption site is fully saturated and the precipitation process takes place.

The flow of effluent along preferred pathways and through large pores was found to reduce P adsorption at equilibrium in two coarse textured soils (Nagpal 1986). Each of the soils (gravelly sandy loam and gravelly loamy sand) was irrigated for 609 days with water having the following P concentrations and hydraulic loading rates: low rate – 7.8 mg/P/l at 158 mm/day; moderate rate – 15.7 mg P/l at 158 mm/day, and high rate – 15.7 mg P/l at 315 mm/day. Doubling the P loading by increasing the P concentration, while maintaining hydraulic loading constant (at 158 mm/day), decreased the P adsorption at equilibrium by 2–11%. On the other hand, the same increase in P loading by elevating the hydraulic loading decreased P adsorption at equilibrium by 12–34%. The loadings rates employed in this experiment were much higher than those normally used in irrigated soils (310 mm/day). Anaerobic conditions can also enhance phosphate movement through soils.

Nagpal (1985) summarized the results of various column tests subjected to near-continuous leaching under saturated and unsaturated conditions as follows: (1) the P added is initially sorbed by the soil; (2) breakthrough of P occurs after the sorbed quantity of P is equivalent to that sorbed by the soil after 200 h of reaction time in a batch test; and (3) the concentration of P in the leachate increases steadily with time after breakthrough and reaches the value of that in the effluent.

The movement of P through the soil has been extensively studied in the field, especially in relation to high-rate effluent application used for recharging groundwater, or in connection with long-term irrigation of agricultural crops. The P added through sewage effluent is removed, sometimes almost completely, by the top layer of the soil. Nevertheless, combinations of high-rate effluent application, especially of P-rich effluent, with soils having low P adsorption capacities may result in increased concentration of P below the root zone of plants. A similar situation occurs in fertilized or manured fields when P movement through the soil is promoted.

The phosphorus balance in Pennsylvania (USA) soils irrigated with sewage effluent for 9–11 years for disposal purposes was studied by Kardos and Hook (1976). The following four cropland sites located on clay loam soil were included in the study: (1) an irrigated cornfield; (2) a control plot of corn; (3) a reed canarygrass field; and (4) a previously cultivated, abandoned field; the fifth study site was a hardwood forest located on a sandy loam soil. The two soils had acid reaction with pH of 5.2 and 4.9, respectively. During the irrigation seasons, 50 mm effluent/week were added. More details on the irrigation schedule and amounts of P added are given in Table 3.10. The P uptake by corn and reed Canary grass was considerable, but most of the added P was retained in the soil. Only a small quantity of P was detected below the 1.2-m layer in the different fields. It was less than 1% of the quantity of P added to corn and reed Canary grass fields; about 1% for the hardwood forest and about 3% for the abandoned field.

The status of inorganic P in the above-mentioned Pennsylvania soils was studied by Sommers et al. (1979). Most of the added P was found in the top layer (0–0.3 m) of the soil, but a significant quantity of P travelled into deeper layers of the sandy loam soil during the experimental period. The equilibrium phosphate concentration was greatly increased in the top layer of the effluent-treated soil. The retention of P was higher in the clay loam soil, since its sesquioxide content was greater than that of the sandy loam soil.

The long-term effects of effluent irrigation on the status of P in soils was studied in a humid region of the USA (Minnesota) by Latterell et al. (1982). Three annual applications of effluent (0, 1260 and 2370 mm/yr) having PO_4-P and total P levels of 6.9 and 7.5 mg P/l, respectively, were tested. Other plots included in the study received comparable amounts of PO_4-P as fertilizer only. The concentrations of organic P in the 0–0.15 m soil layer after 5 years of treatment

Table 3.10 Total amounts of P added by wastewater (applied at the rate of 50 mm/week) or fertilizer, removed in harvested crops or leached past the 1.2-m soil depth during 1963–1973 (Kardos Hook 1976)

Crop	Soil texture	Range of P concentration in effluent	Amounts of P added to soil	P taken up by crop	P leached below 1.2 m depth
		mg P/l	kg P/ha		
Corn, irrigated	Clay loam	3.3–7.2	791	346	7
Corn,[b] control	Clay loam	—	438	162	2
Reed canary grass	Clay loam	4.7–20.6	2215	491	13
Abandoned field		3.1–10.4	1044	—	35[a]
Hardwood	sandy loam	5.5–12.6	1806	—	21[a]

[a] Probably overestimated.
[b] Not irrigated, received conventional fertilizers.

were 178, 282 and 366 mg/kg soil in the 0-, 1260- and 2370-mm effluent treatments, respectively. The available P measured by Bray extract (extraction with 0.03N NH_4F in 0.025N HCl solution) in the 0–0.15, 0.15–0.3, 0.3–0.45 and 0.45–0.6 m soil layers of the nonirrigated treatment were 47, 8, 2 and 3 mg P/kg soil, respectively. The corresponding P values for the 1260- and 2370-mm effluent treatments were 25, 17, 4 and 3, and 85, 42, 12 and 11 mg P/kg, respectively. These data indicate some P penetration into the deeper soil layer in the high-effluent irrigation treatment.

The effect of irrigation with sewage effluent vs fresh water on the level of available P in a clay soil (about 70% clay) was studied under semi-arid Mediterranean conditions in the Yizre'el valley, Israel, from 1976 to 1982 (Feigin, Dag and Giskin, unpublished data). The standard irrigation of cotton, the main test crop, consisted of about 500 mm of sewage effluent or fresh water applied by means of three or four sprinkler irrigations per season. Effluent quantities that were 20% higher and 10% lower than this value were also included. Each irrigation treatment was combined with four fertilizer treatments. Table 3.11 shows the effect of the irrigation treatments on the level of available P in the top layer (0–0.2 m) of soil as determined by standard $NaHCO_3$ extraction (Olsen and Sommers 1982). The slight increase in available P concentration in the standard freshwater treatment from 1979 to 1982 was due to the fertilizer application of 100 kg P/ha given in November 1979. The cumulative quantities of P added to the soil through the standard effluent irrigation treatment were 192, 302, 378 and 507 kg/ha at the end of the 1978, 1979, 1980 and 1982 seasons, respectively. This resulted in a tendency of increasing levels of available P in the soils, although most of the P taken up by cotton was found in the seed and therefore removed from the field. Nevertheless, the values of available P detected in the effluent-

Table 3.11 Effect of irrigation with sewage effluent on the level of available P in the top layer (0–20 cm) of a clay soil (mg P/kg air-dry soil; Feigin, Dag and Giskin, unpublished data)

Date	Fresh water	Sewage effluent		
	Standard irrigation	Standard irrigation	Standard irrigation + 20% in volume	Low frequency
May 1979	6	12	10	13
Nov 1979	5	20	18	12
May 1980	8	15	15	15
Nov 1980	9	20	23	18
Oct 1982	9	33	25	20

treated soil was still in the normal range for many crops, but too low for others (e.g., the recommended values of $NaHCO_3$-extractable P in soils planted with certain vegetable crops are 30 mg P/kg soil). No increase in the level of available P was detected in soil layers deeper than 0.4 m.

Hook (1983) reviewed available data on P movement in wastewater-irrigated soils. Relevant field studies found little evidence of P penetration into subsoil. However, Hook also mentioned technical difficulties that cast some doubt on these observations.

3.4.6 Phosphorus Uptake by Crops

The uptake of P by agricultural crops varies greatly; common values are 10–100 kg ha^{-1} hr^{-1}, with most values being 20–50 kg ha^{-1} yr^{-1}. The total P uptake by crops is lower than that of N and K, its concentration in plant tissues usually being one-tenth to one-fifth of that of N and K. The P taken up may be removed from the field in the harvested portion or may be reincorporated into the soil in the residue. Typical values of P removed from agricultural fields through different plant parts or whole above-ground plant material are shown in Table 3.12. Some of the data for cotton, sugar beet and grasses were obtained from effluent-irrigated crops. Dag et al. (1984) found that the quantities of P taken up by cotton plants (seed or whole plants) during a 5-year study on a clay soil in the Yizre'el Valley (Israel) were considerably greater in effluent-treated plots in comparison with freshwater-irrigated plots (Fig. 3.19). This was the result of both higher P concentration in plant tissue and greater production of dry matter. Nitrogen fertilization was found to influence P uptake in the freshwater treatments.

The quantity of P added through standard irrigation with sewage effluent (usually 3–10 mm/day) seldom results in the addition of P above the normal levels

Table 3.12 Phosphorus removal through different yield components in crops (Broadbent and Reisenauer 1985)[a]

Crop	Typical yield Mg/ha	Quantity of P removed (kg P/ha)	Remarks
Alfalfa	14.5	34.5	
Barley			
Grain	5.0	16.0	
Straw	6.2	7.9	
Corn			
Grain	12.0	35.0	
Silage	5	5.2	
Cotton			
Seed + lint[b]	3.5–6.0	11.0–21.0	Seed + lint + stalk:
seed	2.1	21.0	21.0–52.0 kg P/ha
Stalk	1.6	11.6	
Sorghum			
Grain	5	15.0	
Stover	4.5	7.0	
Soybean			
Grain	3.1	16.9	
Stover	3.1	15.6	
Sugar beet[c]	15.8	38.0–42.0	Sugar yield
Wheat			
Grain	5	7.9	
Stovor	8.8	18.1	
Reed canary grass[d]	56	63	
Rhodes grass[c]	10.7–22.2	37–88	Dry matter
Tomato			
Fruit	62	8.5	
Vines	75	13.5	
	—		
Hardwood forest[d]	—	9	

[a] Except as otherwise noted.
[b] Feigin and Dag (unpubl. data).
[c] Feigin et al. (1979).
[d] Wolcott and Cook (1976).

given by fertilization. For instance, 500 mm/yr of secondary sewage effluent adds 50 kg P/ha^{-1} yr^{-1}, if the P content in the effluent is 10 mg/l. Thus, in semi-arid regions of Israel, where the quantity of irrigation water only seldom exceeds 1000 mm/yr, the amount of P added amounts to 100 kg ha^{-1} yr^{-1}. However, if the quantities of effluent applied are much greater, as in the case of high-rate applications for disposal purposes, the amounts of P added exceed those usually applied through fertilization, and accumulation of available P in soil may cause development of micronutrient deficiencies (Ryden and Pratt 1980).

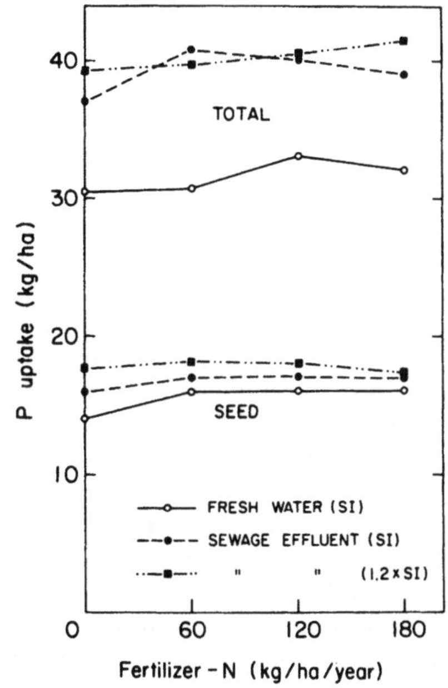

Fig. 3.19 Average uptake of P by the seed and the whole cotton plants irrigated for five seasons with fresh water given as standard irrigation (3 irrigations per season) (*SI*), and sewage effluent given as standard irrigation (*SI*) and 1.2 times the amount given in SI treatment. Each irrigation treatment was combined with four fertilizer-N levels (Feigin and Dag, unpubl. data)

3.4.7 Prediction of P Movement and Leaching Through the Soil

Current models for predicting P removal by soils was reviewed by Ryden and Pratt (1980) and Enfield and Ellis (1983). Ryden and Pratt (1980) concluded that the main limitations of available models are due to the following factors: (1) the great spatial and temporal variabilities in hydraulic conductivity of field soils; (2) similar variability in P sorption capacity; (3) lack of information on the effects of wastewater due to its Al, Fe and Ca contents; and (4) lack of data on the soil acidity-alkinility level. Since P reactions in soils cannot be predicted from measurements of simple soil properties that can be mapped in the field or measured quickly in the laboratory, extensive laboratory work is necessary to obtain the relevant information. Ryden and Pratt (1980) have therefore suggested a simple practical empirical model that can be used successfully to predict P removal from effluent by the soil. This model predicts the longevity of treatment sites with respect to P removal.

The Ryden–Pratt model is governed by the following equation:

$$t = \frac{S_p}{I_p - H_p}, \tag{3.31}$$

where t is time in years for the P front to reach a given soil depth; S_p is the sorption capacity of the volume of soil above this depth, given in kg P/ha, I_p is the input of P in $kg\,ha^{-1}\,yr^{-1}$; and H_p is the P removed in harvested crops in $kg\,P\,ha^{-1}\,yr^{-1}$.

The excess P (added P minus the quantity of P removed by the crops) is assumed to react with the soil. Movement of P from a given horizon to a deeper one occurs after its saturation with P. In this context, saturation is defined as the enrichment of the soil with P to the extent that movement of P with the percolating water is significantly above the background value for the original soil material.

The I_p and H_p values can be obtained by means of simple measurements, but determination of S_p is much more complicated, and is estimated form appropriate laboratory measurements (Ryden and Pratt 1980). The estimation of S_p is based on the establishment of a generalized isotherm for equilibrium P sorption by different soils, where the amount of P sorbed at equilibrium is expressed as a fraction of the overall sorption maximum. The value of this maximum is obtained by shaking the soil with a wide range of P concentrations.

A specific example illustrating the usefulness of the Ryden–Pratt P-balance reaction model is given in Fig. 3.20. The total potential adsorption capacity (S_p) of the relevant ten 0.2-m soil depth increments is 4116 kg P ha^{-1} yr^{-1}, and P removal by the crop (H_p) is 50 kg P ha^{-1} yr^{-1}. This results in a breakthrough of P below a depth of 2 m after a 41-year treatment.

Inaccuracies in the model are due to the invalid assumption of abrupt boundaries between the different soil layers, rather than diffuse boundaries which actually exist. In addition, water movement varies due to variation in effluent application rate, soil texture, infiltration, percolation and evaporation. Nevertheless, the model is useful for preliminary estimation of the longevity of specific disposal sites. More information on this model is given elsewhere (Ryden et al. 1981). Using data obtained in California, Enfield (1981) demonstrated the ability of physical chemistry models to describe P transport in soil.

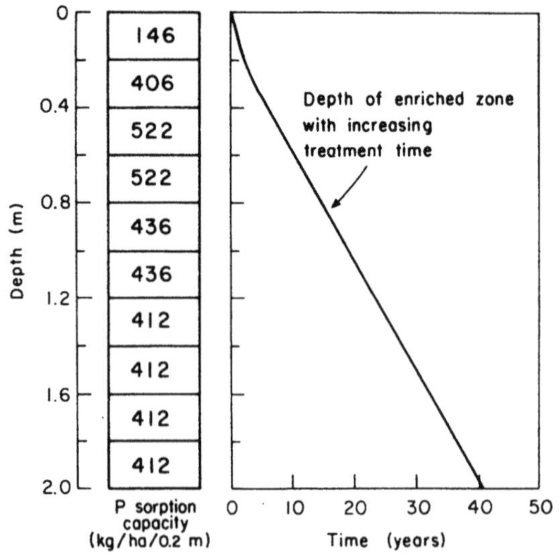

Fig. 3.20 Depth of P-enriched zone with increasing treatment time in a California soil. The P sorption capacity for ten consecutive 0.2-m depth increments is given. The quantities of P added as effluent is 150 kg P ha^{-1} yr^{-1} and that of P removed by crop is 50 kg P ha^{-1} hr^{-1} (After Ryden and Pratt 1980)

Assuming a P concentration in wastewater of 7 mg/l, a crop removal of 30 kg P ha^{-1} yr^{-1} and a 50-year life expectancy of the system, Ellis (1976) gave the following maximum rates for wastewater application to Michigan soils: 1500 mm/yr for silty clay to clay soils, 1400 mm for clay loam, 1350 for loam, 1150 for loamy sand, and 1000/yr mm for sandy loam and sand soils.

3.5 Potassium

3.5.1 Role of Effluent K in Agriculture

Potassium is an essential nutrient element for organisms, and it is taken up by crops in large quantities, similar to, or even greater than those for N. Since the level of available K in the soil is often below that necessary for production of high crop yields, fertilizers and organic materials (e.g., manure) are often added. Furthermore, as K fertilizer is also considered an effective means for obtaining a high-quality product, it is extremely important in high-productive agriculture. The level of K in sewage effluent varies greatly, typical concentrations being 7–20 mg K/l (Table 2.7). Irrigation with such effluent at the standard application rate for cotton in Israel (about 500 mm/season) results in the addition of 35–100 kg ha^{-1} yr^{-1}, which is far below the K uptake by cotton (150–300 kg K ha^{-1} yr^{-1} or greater) or by other crops (see Table 3.13). Consequently, irrigation with sewage effluent often does not satisfy plant needs in K, especially when the level of available K in the soil is low. When higher levels of K are present

Table 3.13 Potassium removal through different yield components of different crops

Crop yield	Typical yield (Mg/ha)	Quantity of K removed (kg/ha)	References
Alfalfa	15	384	Lanyon and Smith (1985)
Red clover	—	80–281	Lanyon and Smith (1985)
Coastal Bermuda grass	17	358	Robinson (1985)
Corn (grain)	5–20	22–85	Welch and Flannery (1985)
(Whole plant)	—	89–395	Welch and Flannery (1985)
Barley	3.2–5.4	81–140	Beaton and Sekhon (1985)
Sorghum (grain)	6.7–11.2	140–233	Beaton and Sekhon (1985)
Peal mellet	—	330	Overman (1975)
Wheat	2.7–6.7	60–189	Beaton and Sekhon (1985)
Oats	3.0–3.6	70–135	Beaton and Sekhon (1985)
Sugar beet (sugar)	13–15.8	396–510	Feigin et al. (1979)
Rhodes grass (dry matter)	21	364	Feigin (unpubl. data)
Cotton (seed cotton)	5.5	200–350	Feigin (unpubl. data)
Tomato (processing)	81	280	Feigin (unpubl. data)
Potato	54–95	230–445	Roberts and McDole (1985)
Rice	9.8	258	DeDatta and Mikkelsen (1985)
Reed canary grass	13.8	280	Wolcott and Cook (1976)

in sewage effluent, it becomes an important source of the element. For instance, the concentration of K in some partially-treated sewage effluents used for irrigation in Israel is 12–28 mg K/l (Feigin et al. 1979) or 20–55 mg K/l (Vaisman et al. 1981). Application of large quantities of such effluent results in the addition of considerable quantities of K to the soil, but K overdose is unusual when standard quantities of sewage effluent are used for irrigation.

The crucial role of potassium in plant, human, and animal nutrition, and the frequent occurrence of K deficiency in crops grown in many soils, have stimulated intensive research on various aspects of soil-plant K. A comprehensive up-to-date review on K in agriculture in a recent international symposium (Munson 1985) and other recent review publications (Mengel 1985; Mengel and Kirkby 1987) deal with the origin of soil K, its transformations in soils and its availability to crops. Specific aspects of K in wastes and wastewater were discussed by Olsen and Barber (1977).

3.5.2 Sources, Phases and Transformations of Soil-K

The level of total K in mineral soils is usually 0.4–29.0 mg/kg (Jackson 1964). The main sources of soil K are aluminosilicates, primary minerals, such as K feldspars, biotite and muscovite micas, secondary minerals, and mica materials, such as illite and its weathering products (e.g., vermiculite; Bertsch and Thomas 1985).

Young soils developed from parent materials rich in K-bearing minerals tend to retain high levels of available K, while highly weathered soils are often K deficient. These differences in the K availability potential and the great variation in soil physicochemical properties, result in the varied behavior in the soil of K derived from fertilizer and in the plant's response to its application.

Structural K, being part of the mineral lattice, is exposed to weathering processes responsible for the formation of less stable K forms. Further weathering of primary and secondary K-bearing aluminosilicate minerals results in the formation of available K.

The most available K is that found in soil solution, and its diffusion to plant roots provides adequate K for plant uptake. The diffusion process depends on soil water content, tortuosity of the diffusion pathway, temperature, diffusion coefficient of K^+ in water and the K^+ concentration gradient (Bertsch and Thomas 1985). The uptake of K^+ by crops is fast, and the level of K^+ in soil solution is relatively low, therefore continuous replenishment of soluble K^+ is necessary to maintain adequate supply to agricultural crops. Exchangeable K, which is readily available to crops, is in direct equilibrium with the soil solution. Equilibria also exist between nonexchangeable and exchangeable K and between both these K forms and the soil solution. Application of K^+ as fertilizer, manure, or wastewater increases the level of K^+ in the soil solution, but in the presence of high-adsorptive clay, much of the added K^+ is adsorbed. In the presence of clay minerals with high fixation capacities, K^+ may become strongly bound and converted to the nonexchangeable forms.

Release of K^+ from soil minerals is by exchange, diffusion and dissolution processes (Feigenbaum et al. 1981; Sparks and Huang 1985). The K level in the soil solution depends on the characteristics of the clay minerals, cation exchange reactions (Ca^{2+} and Na^+ enhance K^+ release) and root activity.

3.5.3 Potassium Uptake by Crops

The potassium percentage in plant tissue is relatively high, which indicates a high crop uptake. Table 3.13 shows that the quantities of K removed from the soil are considerable. Also, the amounts of K taken up by crops are often greater than those added through fertilizer or effluent. Feigin et al. (1979) measured the quantities of K taken up by sugar beet, Rhodes grass and cotton grown in effluent-irrigated soils. The amounts of K removed by sugar beet and Rhodes grass were four and two times greater, respectively, than those added through the effluent. The uptake by cotton plants was also two to three times greater than applied, but most of it was later incorporated into the soil with the vegetative parts of the crop.

Palazzo (1977) noted a reduction in the level of exchangeable K in soil following irrigation of forage crops at rates of 50, 85 and 250 mm/week with wastewater having a concentration of 8.8 mg K/l; this supplied 114, 172 and 306 kg K ha^{-1} yr^{-1}, respectively. The reduction was partially related to plant K uptake, which was on the average 247, 244 and 346 kg K ha^{-1} yr^{-1}, respectively. The level of K in the crop tissue was also reduced; it was 26.7, 23.7 and 27.2 g K/kg dry matter, respectively, in the first growing season, and 21.0, 22.2 and 19.4 g K/kg in the second. These data indicate that the relevant forage crops (a mixture of reed canary grass (*Phalaris arundiacea* L.) timothy grass (*Phelum pratense* L.) and smooth bromegrass (*Bromus inermis* Leyss.) required more K than that added through the effluent.

3.5.4 Potassium Transport in Soils

Potassium transport in soils is of both economic and environmental importance. The K in the soil solution is available for plant uptake, reactions with other soil constituents or leaching (Sparks and Huang 1985). The retaining of added K depends to a great extent on the CEC of the soil. The greater the CEC, the less K moves through the soil profile. Coarse-textured soils allow K leaching, while soils rich in clay retain K. Feigin and Dag (unpublished data) noted that the available K in a clay soil (above 70% clay) in Israel was not considerably changed after 7 years of irrigation with sewage effluent, a period during which up to 870 kg K/ha was added. No K increase was detected below the plowed layer (0–0.2 m).

Increasing the pH in sandy, acid soils to 6–6.5 reduced K leaching in the eastern coast of the USA (Sparks and Huang 1985) due to the presence of Al^{3+} in place of Ca^{2+} in the acid sand solution.

Bertsch and Thomas (1985) reviewed K leaching and presented data collected under different soil and climatic conditions showing K losses by leaching to range between 1.5 and 57 $kg\,ha^{-1}\,yr^{-1}$. They found that cropping of sandy soils resulted in a great reduction (80–90%) in K leaching. Upward movement of K salts may result from evapotranspiration (Gammon 1957). Timing, amount and intensity of rainfall are also important factors affecting K leaching.

Potassium losses by soil erosion and runoff are also significant (Black 1968).

3.6 Trace Elements

3.6.1 Introduction

Trace elements are usually present in low concentrations in natural systems. Certain of these chemicals are especially important in waste-treated soils. Much of the information on trace elements originates from soil-plant systems treated with sewage sludge, which are the source of many of the data presented and discussed below. Since both sludge and effluent are derived from the same source (raw sewage), carefully selected data can be successfully adopted for fields irrigated with sewage effluent.

Page (1974) mentioned the following trace elements as important components of sewage sludge: silver (Ag), arsenic (As), boron (B), barium (Ba), cadmium (Cd), cobalt (Co), chromium (Cr), copper (Cu), mercury (Hg), manganese (Mn), molybdenum (Mo), nickel (Ni), lead (Pb), selenium (Se), tin (Sn), vanadium (V). and zinc (Zn). Certain trace elements, including most of those listed by Page, are heavy metals, with a density greater than 5 $g\,cm^{-3}$, although by common usage other metals are also included in this list (Tiller 1989). The following 12 out of the 38 existing heavy metals are used in industry and are therefore important potential pollutants: Cd, Cr, Co, Cu, Fe, Hg, Mn, Mo, Ni, Pb, Sn and Zn (Jones and Jarvis 1981). Tiller (1989) lists Cd, Pb, Cu, Zn, Ni, Sb and Bi as most important environmental pollutants.

Trace elements participate in many of the biochemical processes that greatly affect plants, animals and humans. The following trace elements are essential for plant development and growth (and are therefore named micronutrients): Fe, Mn, Zn, Cu, Mo and B. Those required by animals are Mn, Cu, Zn, Fe, Mo, Se, I, Co, F, Cr, Ni and V. Sodium, Cl, Co, V and Ni are also mentioned as being essential for certain plants (Tiller 1989). On the other hand, other trace elements (Cd, Hg, Pb included) have been shown to be nonessential to plants and animals. Both the biologically required and the other relevant trace elements become toxic to organising when present at concentrations exceeding certain threshold levels of their available forms. The margin between recommended (or acceptable) and toxic concentrations is often quite narrow, as in the case of B. The health hazard posed by excessive levels of trace elements in soils, and consequently in the tissues of plants, animals and humans consuming the plant materials, is therefore of great public concern (Allaway 1977; Purvis 1985). Movement of trace elements to the ground water is also considered as a serious threat under these conditions.

The quantities of trace elements added to the soil depend on the origin of the wastes and wastewater used. The concentration of trace elements in sewage water varies greatly (Thomas and Law 1977; Page et al. 1981). High concentrations are found in industrial sewage effluent, especially in raw sewage, while the levels in domestic effluents and in advanced-treated sewage effluent are very low and negligible, respectively. Data on the range and typical levels of trace elements in sewage effluent are given in Table 2.2.

Some trace elements are hazardous to organisms, while the toxicity of others is low. This is the result of their biochemical effect on living organisms, their physicochemical characteristics, and the forms and actual quantities added to specific soil-plant combinations. Cadmium, Cu, Mo, Ni and Zn are considered more potentially toxic than other elements applied to soils treated with waste or wastewater. For various reasons, including their presence at low level in wastewater and the availability of suitable agronomic management practices, Mn, Fe, Al, Cr, As, Se, Sb, Pb and Hg are categorized as posing little environmental hazard (Page et al. 1981). Boron is also mentioned as a potential hazard if added in an uncontrolled manner (see also Sect. 3.1.). A discussion of the role of trace elements in agricultural systems is given by Mordtvedt et al. (1972).

3.6.2 Practical Considerations

As noted above, the quantity of trace elements added to soils by means of sewage effluent irrigation is generally low. Bouwer and Chaney (1974) concluded that, at applications rates to avoid nitrate pollution, greater amounts of heavy metals can be added to the soil in one year with sludge than may be added in a century of effluent irrigation. Consequently, most research on trace elements in soils treated with waste deals with sludge application rather than with effluent irrigation. Further information is derived from studies on plant nutrition and from research on industrial trace element pollution through aerosols, mining wastes and the use of industrial and agricultural chemcials (Jones and Jarvis 1981; Purvis 1985; Tiller 1989).

Despite their low content, the level of trace elements is an important characteristic of secondary effluent, due to its extensive use for irrigation and groundwater recharge (Thomas and Law 1977). According to Page and Chang (1985), although it may take a century before trace element levels in wastewater-irrigated soils reach the currently proposed upper limit, the potentially harmful effects in the future should not be ignored. Since most of the trace element material tends to accumulate in the soil, long-term irrigation may substantially increase their levels in treated fields. Furthermore, owing to the great differences among effluents, information on the trace element levels in any effluent used for irrigation must be obtained and their levels in soil and plant tissues monitored. In addition, basic information on trace element-soil-plant relations should be consulted. The movement of trace elements in soil profiles has received considerable attention, since even a slow transport through soil and subsoil material may result in an increased content of trace elements in the groundwater.

The following sections will present relevant information on soil and plant processes and specific effects of trace elements on animals and humans.

3.6.3 Trace Element Reactions in Soil

Trace elements are found in sewage effluent both in the suspended solids and in the liquid fractions. Those associated with the suspended solids accumulate on the soil surface as the effluent infiltrates into the soil, while those dissolved in the liquid component penetrate into the soil. The trace elements derived from both fractions interact with soil components. Tillage operations taking place in agricultural fields tend to homogenize the plow layer into which all the trace elements found in the effluent are added. The trace elements reaching soil-plant systems participate in different soil reactions, the principal ones being ion exchange, precipitation, surface adsorption and organic complexing. Prediction of trace element behavior in soils must be based on a knowledge of each of these reactions (Page et al. 1981). A schematic description of trace element transformations in soil-plant systems is given in Fig. 3.21.

Many studies have demonstrated the high capacity of soils to retain trace elements. Jones and Jarvis (1981) have noted the low level of soluble trace elements found in soils, which is usually below that predicted by the solubility product principle, and attribute the low solubility to exchange and specific

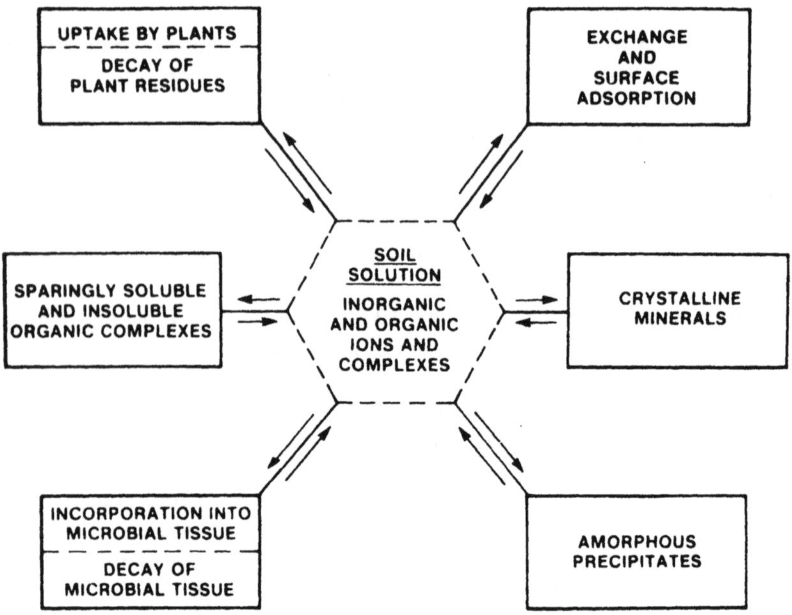

Fig. 3.21 Possible pathways of trace elements in soils (Page et al. 1981)

absorption. McBride (1989) considered the following processes as determining the solubility of heavy metals in the soil: (1) ion exchange on layer silicate; (2) chemisorption on mineral surfaces; (3) reduction, precipitation and solid solutions; (4) redox processes, taking into consideration oxidation of metals, metal oxides and dissolution of metals by organic materials; (5) metal adsorption by organic matter; and (6) speciation. The extensive analysis of the available data was followed by the conclusion that "accounting for all processes that could determine the solubility of trace metals in soils is beyond the capability of any single theoretical approach or model."

The soil pH is also a determinative factor for trace element solubility in soils. Page and Chang (1985) report that in general the solubility of cationic trace element species increases and that of anionic ones decreases with decreasing pH. Small changes in soil pH can result in major differences in the absorption of trace elements on the soil surface (Jones and Jarvis 1981). Hydrous oxides of Mn and Fe were found to be responsible for the reduction of certain heavy metals in the soil.

The presence of organic matter in the soil greatly affects the solubility of trace elements (Stevenson and Ardakani 1972; McBride 1989). Williams et al. (1980) mentioned organic acids, amino acids and fulvic acid, as well as biological systems and biological residues, as influencing trace element solubility in soils. The addition of organic matter was often found to decrease the sorption of trace elements on the soil. The reduced Cd sorption in the presence of sewage sludge was explained by preferential association of Cd with aqueous solution components (Neal and Sposito 1986). The great influence of chelates on the solubility and availability of trace metals in soils has often been reported (Norvell 1972).

Many greenhouse and field studies have been carried out during the last two decades to elucidate the effect of sewage sludge, as well as other wastes, on the level of trace elements in soils. Korte et al. (1976) found that the most useful information for predicting soil effectiveness for trace element retention relates to soil texture, soil surface area, and the content of hydrous oxides and free lime. Soil clay has often been considered as a major factor determining trace element solubility (Williams et al. 1980). The effect of pH is also often mentioned as a predominating factor (Chaney and Giordano 1977). Many field studies have proven that, in general, many metal elements are much more soluble under acid conditions and their solubility declines on addition of alkaline materials (such as lime). Data on the solubility of trace elements in soils under equilibrium conditions is provided by Lindsay (1979).

The movement of trace elements in soils is usually very low, and most of the added material accumulates in the soil layer into which the relevant waste materials are incorporated. Williams et al. (1980) reviewed relevant literature and noted that in most cases the trace elements added through different wastes were either retained in the top soil layer or were moved a few centimeters below the treated layer. A 9-year investigation in Oakland, California by Williams et al. (1980, 1984, 1985, 1987) studied metal movement in 2 different sludge-treated soils, having an initial pH of 4.5 and 7.4, respectively. The movement of metal trace elements after 8 consecutive applications of up to 225 mg ha^{-1} yr^{-1}

was limited in both soils. The total Cd, Cu, Pb, Zn, Cr and Hg levels in the treated top soil (0–20 cm) greatly increased with time and addition rate, but almost no movement below this layer was observed over the 9-year study period.

Dowdy and Volk (1983) concluded that, although some studies have noted movement of trace metals to layers below the soil surface or depth of incorporation, it appears that only Zn has the potential to move in soils. Movement of heavy metals in soils will occur in sandy, acid, low-organic matter soils, subjected to heavy rainfall or irrigation. Open channels or cracks increase trace elements movement in soils, but elsewhere such movement is usually limited.

Important information on trace element accumulation and movement in soil profiles was obtained from the long-term Werribee experiment in Australia (Evans et al. 1979). Most of the elements accumulated in the top layer of a clay loam soil irrigated for 80 years with sewage effluent, having (in 1979) the following trace element content (in mg/l units): Cd 0.015; Cu 0.35; Cr 0.40; Pb 0.30; Ni 0.15; and Zn 0.8 (Fig. 3.22). The quantities of trace elements added in 1979 were in most cases lower than those recommended for long-term irrigation (no data were available for previous years). Despite the considerable increase in trace elements in the soil, Evans et al. state that except for Pb and Zn, the levels remain within the normal range. The level of trace elements in plant material increased, but the authors did not detect a significant increase in heavy metals in the liver and kidneys of grazing animals.

King (1982) studied the effect of land application of treated vs. untreated industrial wastewater of fescue grass (*Festuca arundinacea* Schreb) growing on a kaolinitic clay soil. Effluent was applied for 2 years at a rate of 1.4 cm/week, and for a subsequent 2 years at 3 treatment rates of 1.5, 2.0 and 3.0 cm of effluent/week. The pH of the soil was increased to 7.4–7.8. Adhesion of effluent to the plants was evidenced by a higher concentration of Mn in the plants, but no such increase was detected in washed plants. The concentration of Ni, Pb and Cr in the unwashed material was slightly increased but was below concentrations considered toxic to livestock. In a recent publication, Schirado et al. (1986) presented data indicating movement in soil of several heavy metals originated in a domestic and industrial raw sewage. The study area is located near Santiago, Chile, with a mean annual rainfall of 340 mm. Analytical determination proved that Cd, Cr, Cu, Ni and Zn migrate from the cultivated soil layers into deeper layers.

The effect of irrigation with treated sewage effluent was studied in a pasture near Christchurch, New Zealand (Quin and Syers 1978). The metal levels in the effluent of domestic origin was low (the concentration of Cd was < 0.0001 mg/l, that of Cr, Ni and Co < 0.001, that of Cu and Pb 0.001 and that of Zn and Mn 0.015 mg/l). The levels of 0.1 N HCl extractable metals (Zn, Cu, Co and Mn) in the soil were slightly higher after 16 years of irrigation, but their content in the pasture and the Cd/Zn ratio were not affected. Quin and Syers concluded that the effluent application of 840 mm effluent/yr can be continued indefinitely without causing a hazardous heavy metal buildup.

Page and Chang (1985) calculated the annual inputs of important trace elements added to the soil by irrigation and the length of time needed to reach

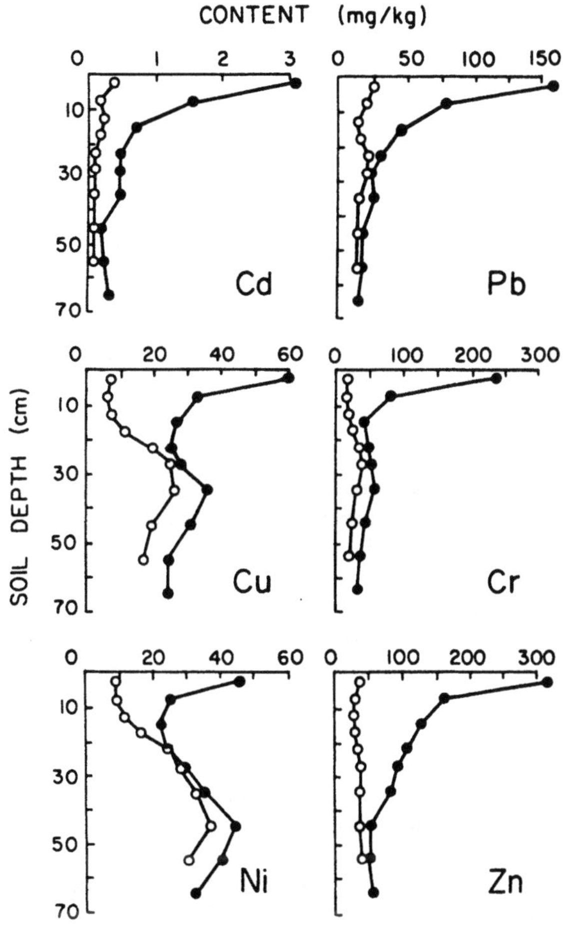

Fig. 3.22 Total (●) and 0.1 N HCl extractable (○) trace element content of a soil profile from the long-term Werribee experiment (Evans et al. 1979)

soil loading limits. Their calculations were based on an annual application of 1200 mm effluent and on a typical concentration of the relevant elements in the effluent. In the absence of trace element input criteria for effluents, the US EPA proposed (see Page and Chang 1985) that the limit for sludge application be taken as a reference and the US EPA recommendation for trace element application for soils having different cation exchange capacities were adopted. In Israel, the highest annual application of effluents is given to Rhodes grass (about 1000 mm irrigation water/yr, in addition to 400–600 mm rainfall in the main growing regions), and the aforementioned approach by Page and Chang (1985) was adopted for these conditions (Table 3.14). Since cotton, a major sewage-effluent consumer, receives half the amount of irrigation water given to Rhodes grass, twice the time period is needed to reach the same loading of

Table 3.14 Calculated length of time for a wastewater irrigated Rhodes grass field to reach trace element loading limits in soils having different cation exchange capacity (After Page and Chang 1985)

Element	Typical Concentration (mg/l)	Annual input in 1000 mm of water (kg/ha)	Suggested loading kg/ha at soil CEC[a]			Time in years to reach soil loading limit at CEC		
			<5	5–15	>15	<5	5–15	>15
Cadmium	0.005	0.05	5	10	20	100	200	400
Copper	0.1	1.0	125	250	500	125	250	500
Nickel	0.02	0.2	125	250	500	625	1250	2500
Zinc	0.15	1.5	250	500	1000	165	330	660
Lead	0.05	0.5	500	1000	2000	1000	2000	4000

[a]CEC is given in mmol(+)/100 g soil.

trace elements. Table 3.14 shows that trace elements are added at relatively low levels and the number of years needed for reaching the specific loading limits are hundreds in the case of Cd, Cu, Ni, and Zn, and thousands in the case of Ni and Pb.

A relatively small amount of trace elements applied through a typical slow rate wastewater application is removed by crops (Page et al. 1981). For a forage crop the following levels (in %) have been obtained: As 5; B 1.25; Cd 2.5; Cr 0.5; Cu 3.75; Hg 0.55; Mo 5; Ni 6.25; Pb 1; Se 2.5; Zn 8.3.

3.6.4 Beneficial and Hazardous Effects of Trace Elements on Plants and Animals

The addition of trace elements by means of effluent irrigation can be beneficial under certain circumstances. Thus, the micronutrient deficiencies occurring under various soil-plant conditions, especially in calcareous soils, can be corrected by adding mineral or chelated trace element substances. Since some trace elements are required by animals, their addition through plants irrigated with sewage effluent can be considered as another advantage obtained from wastewater use (in addition to the contribution of water and plant nutrients). However, the main concern today is the danger of accumulation of excessive levels of trace elements leading to toxicity and hence to health and environmental hazards. Such problems are not restricted to wastes use, but also occur in heavily industrialized areas as a result of other contamination factors (Purvis 1985).

Much research work has been carried out to obtain reliable information on trace element uptake and accumulation in plants grown in effluent or sludge-treated soils (Page 1974; Bouwer and Chaney 1974; CAST 1976; Chaney and Giordano 1977; Allaway 1977; Chang et al. 1981; Page et al. 1981; Logan and Chaney 1983; Purvis 1985; Tiller 1989; Chaney 1989). There is ample evidence showing that trace elements may accumulate in plant tissues, but plant properties differ greatly and the effect of soil conditions is often decisive. Furthermore, the distribution of trace elements within plant organs also varies greatly among plant species. Since in most cases accumulation of trace elements in effluent-irrigated soil, especially with advanced-treated effluent, is slow and small, no serious accumulation in plant tissues is anticipated after short irrigation periods. On the other hand, some effluent originating from industrial sources, and particularly sewage sludge, may contribute large quantities of trace elements.

In addition to the information on plant response to trace elements and their accumulation in plant tissues obtained for raw sewage, or sludge amended soils, laboratory studies have been carried out on solutions and simulated effluent rich in trace elements. Accumulation of trace elements in plant tissue may detrimentally affect crop growth, but some trace metals accumulate in plant material without any external signs. Consumption of such materials through the food chain may adversely affect animals and humans. Much attention has been paid to Cd and Zn due to toxicity hazard and their potential for accumulating in plant tissue (Chang et al. 1982). Other elements considered as most hazardous due to their occurrence in large quantities in wastes such as sewage sludge and their toxicity affects are B, Cu, Mo and Ni (CAST 1976).

Many reports have shown that the presence of high levels of trace elements in the soil may result in a similar situation in plant tissues. In other cases, the levels of trace metals in crop constituents remained low, even though large quantities of wastes rich in trace elements were added to the soil. This discrepancy results from differences in soil conditions (e.g., pH) and the properties of plants grown. Some typical examples are given below.

Vlamis et al. (1985) found that the level of Cd and Zn in barley plants was affected by the level of these trace elements in the sewage sludge added to the acid soil. In their experiment, sludge A was richer in trace metals than sludge

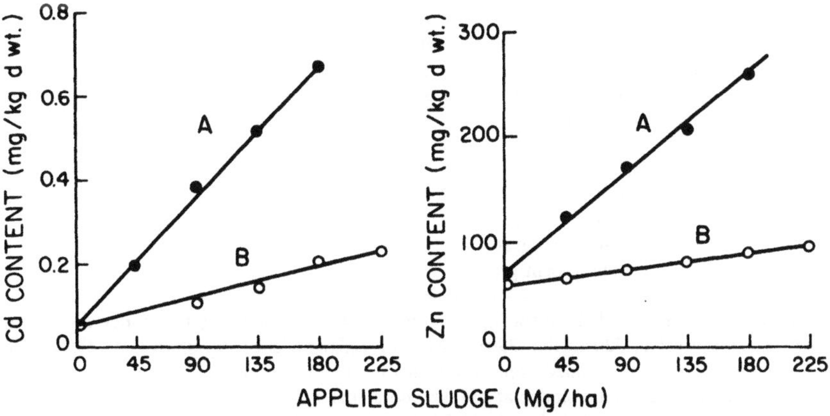

Fig. 3.23 Level of Cd and Zn in the grain of barley plants grown for 7 years in an acid soil treated with two different sludges (*A* and *B*, see text; after Vlamis et al. 1985)

B, and this was reflected in the increased levels in the sludge-A treated barley grain (Fig. 3.23). The yield of barley was positively affected by the sludge. The authors concluded that the application of sludge B, having low levels of Cd and Zn, and 45 Mg N/ha, for 7 years did not increase the trace metal contents to prohibited levels, and that the period could be extended multifold. On the other hand, sludge A, together with greater amounts of N containing a higher proportion of NH_4, both reduced pH and increased the contribution of Cd and Zn to the soil. Under such conditions, the uptake of Zn and Cd by plants was much greater than in the case of sludge B. Even in the case of sludge A, the concentration of Cd and Zn were acceptable after the addition of sludge in agronomic quantities. Similar effects were found for Zn and Cu in snap beans (Dowdy et al. 1978), and the levels of Cd and Zn in the edible tissue of various vegetable crops were slightly, although significantly elevated (Dowdy and Larson 1975). In some cases, high concentrations of trace metals in different plants resulted in lower yields (Chang et al. 1981), whereas no adverse effect on the yield was

Table 3.15 Relative sensitivity of crops to sludge-applied heavy metals[a] (Logan and Chaney 1983)

Very sensitive[b]	Sensitive[c]	Tolerant[d]	Very tolerant[e]
Chard	Mustard	Cauliflower	Corn
Lettuce	Kale	Cucumber	Sudan grass
Red beet	Spinach	Zucchini squash	Smooth bromegrass
Carrot	Broccoli		'Merlin' red fescue
Turnip	Radish	Flat pea	
Peanut	Tomato		
	Marigold	Oat	
Ladino clover		Orchard grass	
Alsike clover	Zigzag, red, Kura and crimson clover	Japanese bromegrass	
Crown vetch	Alfalfa	Switchgrass	
'Arc' alfalfa	Korean lespedeza	Red top	
White sweet clover	Sericea lespedeza	Buffel grass	
Yellow sweet clover	Blue lupin	Tall fescue	
	Birdsfoot trefoil	Red fescue	
Weeping love grass	Hairy vetch	Kentucky bluegrass	
Lehman love grass	Soybean		
Deer tongue	Snapbean		
	Timothy		
	Colonial bent grass		
	Perennial ryegrass		
	Creeping bent grass		

[a]Sassafras sandy loam amended with a highly stabilized and leached digested sludge containing 5300 mg Zn; 2400 mg Cu; 320 mg Ni; 390 mg Mn; and 23 mg Cd/kg dry sludge. At 5% sludge, maximum cumulative recommended applications of Zn and Cu are made.
[b]Injured at 10% of a high metal sludge at pH 6.5 and at pH 5.5.
[c]Injured at 10% of a high metal sludge at pH 5.5, but not at pH 6.5.
[d]Injured at 25% high metal sludge at pH 5.5, but not at pH 6.5, and not at 10% sludge at pH 5.5 or 6.5.
[e]Not injured even at 25% sludge, pH 5.5.

reported in other cases (e.g., Rappaport et al. 1987). The level of Cd and Zn in crop tissue has been shown to be linearly related to the quantity of sludge added to soil (e.g., Fig. 3.23). Recent evidence, however, indicated the development of a plateau in the response curve of lettuce grown in a soil exposed to long-term sludge application (Chaney 1989).

Logan and Chaney (1983) presented the relative sensitivity of many plants to sludge application (Table 3.15). A highly stabilized and leached digested sludge, with different levels of Zn, Cu, Ni, Mn and Cd, has been added to a sandy loam soil. The sensitivity level is based on crop response to the heavy metal application under different soil pH.

3.6.5 Food Chain Aspects

Apart from the adverse effect certain elements may have on crop yield, accumulation of such materials in plant tissue may cause an increase in the trace element level in the food chain.

Food chain aspects related to the use of sewage effluent and sludge have been frequently discussed (e.g., CAST 1976; Allaway 1977; Underwood 1979; Jones and Jarvis 1981; Purvis 1985; Chaney 1989). Phytotoxicity is an important barrier to the use of trace element-enriched plant material for food. Some

Table 3.16 Range in concentrations of selected trace elements in dry digested sewage sludges[a] (Chaney 1989)

Element[b]	Reported range Minimum	Maximum	Typical median sludge	Typical soil	Maximum domestic sludge
As	1.1	230	10	—	—
Cd	1	3410	10	0.1	25
Cd/Zn, %	0.1	110	0.8	—	1.5
Co	11.3	2490	30	—	200
Cu	84	17000	800	15	1000
Cr	10	99000	500	25	1000
F	80	33500	260	200	1000
Fe, %	0.1	15.4	1.7	2.0	4.0
Hg	0.6	56	6	—	10
Mn	32	9870	260	500	—
Mo	0.1	3700	20	—	35
Ni	2	5300	80	25	200
Pb[c]	13	26000	250	15	500
Sn	2.6	329	14	—	—
Se	1.7	17.2	5	—	—
Zn	101	49000	1700	50	2500

[a]Composting using wood chips as a bulking agent generally produces composted sludge 50% as high in trace elements as a digested sludge from the same treatment plant.
[b]mg/kg unless otherwise noted.
[c]Sludge Pb concentration has dropped significantly during the 1980's in the USA due to reduction of leaded gasoline.

elements affect plant growth at a level much lower than that toxic to animals and humans. Consequently, even though an element is taken up by a plant, and is translocated into the foliage, its level in the tissue is limited by phytotoxicity (Chaney 1989). The "Soil-Plant-Barrier" (Chaney 1989) prevents a further accumulation of trace elements in plant tissue, and the concentration of toxic elements in plant parts consumed by humans or animals does not exceed the permissible levels. Table 3.16 presents specific information concerning 15 trace elements, and Table 3.17 shows that phytotoxicity limits the levels of at least eight elements in this list. Specific food chain aspects will be discussed within the framework of Sect. 3.6.6.

3.6.6 Effects of Specific Trace Elements on Soil, Plant, and Food Chain

Arsenic (As). Recorded cases of arsenic pollution have usually been attributed to pesticide application or to the use of industrial products. Most of the information on the potential toxicity of As is obtained from studies carried out in highly contaminated sites (Page et al. 1981). The range and median levels of arsenic in dry digested sludge are 1.1–230 and 10 mg/kg (Table 3.16; in comparison with < 0.005–0.023 mg/l in municipal secondary effluents). Toxicity may occur after application of 20 Mg/ha of high-As sludge. In the soil, the element reverts to arsenate which is strongly held by clay; the availability of As is reduced in fine-textured soils.

Arsenic usually accumulates in the plant roots, so that its content in the edible plant parts is generally below the critical level of 3.0 mg/kg (Table 3.17). The typical As levels in plant tissue are between 0.1 and 5 mg/kg dry matter (Page et al. 1981), and in soil it is 6 mg/kg (Page et al. 1981). Its level in forage crop is 0.01–1 mg/kg with toxicity occurring at 3–10 mg/kg (Table 3.17). The estimated concentration in soil solution at toxic levels, estimated at 30 times its concentration in seawater (Bohn et al. 1985), is 0.1 mg/l. This is also the recommended maximum concentration in irrigation water (Ayers and Westcot 1985). Toxic effect on crops varies greatly, e.g., 0.05 mg/l is toxic for rice and 12 mg/l for Sudan grass. Toxicity and yield failure precede the production of crops containing hazardous levels in edible parts. Consequently, the soil-plant barrier protects the food chain against As (Chaney 1989). Arsenic is usually added to poultry diet.

Boron (B). Boron is essential for plants, but is not required by animals. About 0.5 mg/l is usually added in nutrient solutions used to nurture plants in soilless cultures. Boron is associated with salinity problems and its presence in arid soils is widespread. Since its main form in wastewater is probably nondissociated boric acid, $B(OH)_3$ (an uncharged molecule) boron moves faster in soil in comparison with other trace elements (Page et al. 1981). It can be leached out, although leaching requires large quantities of water. Its typical concentration in crop tissue is 5–30 mg/kg (Page et al. 1981). Details on B toxicity levels are presented and discussed in Sect. 3.1.

Table 3.17 Maximum tolerable levels of dietary minerals for domestic livestock in comparison with levels in forages (Chaney 1989)

Element	Soil-plant Barrier	Level in plant foliage		Maximum levels chronically tolerated[a]			
		Normal	Phytotoxic	Cattle	Sheep	Swine	Chicken
		(mg/kg dry foliage)		(mg/kg dry diet)			
As_{inorg}	Yes	0.01–1	3–10	50	50	50	50
B	Yes	7–75	75	150	(150)	(150)	(150)
Cd[b]	Fails	0.1–1	5–700	0.5	0.5	0.5	0.5
Cr^{3+}	Yes	0.1–1	20	(3000)	(3000)	(3000)	3000
Co	Fails	0.01–0.3	25–100	10	10	10	10
Cu	Yes	3–20	25–40	100	25	250	300
F	Yes?	1–5	—	40	60	150	200
Fe^{2+}	Yes	30–300	—	1000	500	3000	1000
Mn	?	15–150	400–2000	1000	1000	400	2000
Mo	Fails	0.1–3.0	100	10	10	20	100
Ni	Yes	0.1–5	50–100	50	(50)	(100)	(300)
Pb[b]	Yes	2–5	—	30	30	30	30
Se	Fails	0.1–2	100	(2)	(2)	2	2
V	Yes?	0.1–1	10	50	50	(10)	10
Zn	Yes	15–150	500–1500	500	300	1000	1000

[a] Based on NRC (1980). Continuous long-term feeding of minerals at the maximum tolerable levels may cause adverse effects. Levels in parenthesis were estimated (by NRC) by extrapolating between animal species.
[b] Maximum levels tolerated were based on Cd or Pb in liver, kidney, and bone in foods for humans rather than simple tolerance by the animals.

Cadmium (Cd). Cadmium is considered as a potentially serious environmental and health hazard. This is because of its mobility in the food chain and to its high toxicity to plants, animals and humans. The incidence of the Itai-Itai disease and other health problems are associated with the presence of Cd in the food chain (Purvis 1985). The Itai-Itai disease occurred under special conditions in Japan where the water used to irrigate rice originated in a Cd-mining area.

The Cd level in soils varies greatly from 0.01–7.0 mg/kg, the typical concentration being 0.06 mg/kg (Page et al. 1981). The typical Cd concentration in the soil solution is estimated as 0.001 mg/l; in secondary municipal effluents it is <0.005–0.15 mg/l. The recommended upper level in livestock drinking water is 0.05 mg/l (Ayers and Westcot 1985). In sludge, the Cd content is about 1–3000 mg/kg with a median value of about 10 mg/kg. The typical concentration of Cd in plant tissues is 0.1–1.0 mg/kg (Table 3.17).

Bingham et al. (1975, 1976) studied the relation between the levels of available Cd in the soil and in plant tissue. The most sensitive crops, of those tested, (i.e., 50% yield reduction as a result of 10–50 mg Cd/kg soil) were: spinach (*Spinacia oleracea* L.) soybean (*Glycine max* Merr.), curly cress (*Lepidium sativum* L.), romaine lettuce (*Lactuca sativa* L.), sweet corn (*Zea mays* L.), and upland rice (*Oryza sativa* L.). The second, more tolerant, group of crops (50% yield reduction as a result of 50–100 Cd mg/kg soil) included: Sudan grass (*Sorghum halepense* Pers.) carrot (*Daucus carota* L. *var sativa* D.C), field bean (*Phaseolus vulgaris* L.), and wheat (*Triticum aestivum* L.). The third group crops (50% yield reduction as a result of 100–300 mg Cd/kg soil) included; white clover (*Trifolium repens* L.), table beet (*Beta vulgaris,* L.), alfalfa (*Medicago sativa* L.), radish (*Raphanus sativus* L.), and zucchini squash (*Cucurbita pepo* L.). The most tolerant crops, (50% yield reduction as a result of > 300 mg Cd/kg soil) consisted of: Swiss chard (*Beta vulgaris* L. *var Chicla* L.), tall fescue (*Fastuca elatior* L.), cabbage (*Brassica oleracea* L. *var capitata* L.), Bermuda grass (*Cynodon dactylon* Persc.), tomato (*Lycopersicon esculentum* Mil.), and paddy rice (*Oryza sativa* L.).

Even though the aforementioned data was obtained in only one soil, the response in yield to different levels of Cd in the soil and plant tissue varied greatly. The Cd levels in different plant parts differ greatly; for instance, it is 6–30 times greater in the leaves than in the grain of wheat. In other crops (soybeans, oats and sorghum) the grain Cd was 30–100% of that found in the leaves.

The Cd level in animal tissue increases as its level in the diet increases. Cadmium levels in kidneys and liver exceed that in muscles. The quantity of sludge allowed for application is based on predicted Cd intake. Different values suggested for the protection of public health were suggested by the USEPA, WHO and the CEC. This subject has been reviewed recently by Chaney (1989), who suggested a revision in the present limits.

Cobalt (*Co*). Cobalt is required by leguminous plants (< 0.2 mg/kg) and by animals. Its concentration in the soil is 1–40 mg/kg, the typical value being 8 mg/kg (Page et al. 1981). The normal range in plant tissue is 0.1–0.3 mg/kg (Table 3.17). The estimated level in soil solution under natural conditions is 0.01 mg/l, and its recommended upper level for livestock in drinking water is 1.0 mg/l. The higher acceptable level for long-term irrigation is 0.05 mg/l. Cobalt is toxic to tomato plants at 0.1 mg/l in the nutrient solution, and its toxicity level in plant foliage is 25–100 mg/kg dry matter.

Chromium (*Cr*). Chromium is not generally considered as an essential element, but deficiency in humans and animals has been reported. Its toxicity hazard is low. Its level in soil ranges from 5–3000 mg/kg, with a typical concentration of 100 mg/kg (Page et al. 1981). Its estimated level in soil solution is 0.01 mg/l. The range of normal level of Cr^{3+} in plant tissue is 0.1–1 mg/kg and phytotoxicity is found at 20 mg/kg in plant foliage (Chaney 1989). Hexavalent Cr is toxic to plants, but its levels in sludge and natural systems is low to very low. In the soil this element reaches a Cr^{3+} state with low solubility. Due to the limited possibility of hexavalent Cr being present in the soil, Cr does not pose a real hazard to the food chain. The level of Cr in sludge varies greatly. Its estimated level in the soil solution is 0.01 mg/l. Most crops take up small quantities of Cr, but some species can take up considerable quantities; Keefer et al. (1986) found 2 mg Cr/kg or more in the tops of radish plants, but levels as high as

30 mg/kg have also been reported. A restricted translocation from leaves to grain was detected in corn. Plants can tolerate high levels of Cr in soil.

Copper (Cu). Copper is required by both plants and animals, but is also considered as one of the elements posing a potentially serious hazard when present in excessive levels, as in the case of certain sludge-treated soils.

The range of Cu in soils is 2–100 mg/kg with a typical level of 20 mg/kg (Page et al. 1981). In soil solution its level is 0.03–0.3 mg/l. The level of Cu in sewage sludge is 80–17 000 mg/kg with a median value of 800 mg/kg.

Copper is added to nutrient solutions used for soilless cultures at a concentration of about 0.02 mg/l, but it may become toxic to some plants at levels between 0.1 and 1.0 mg/l in such solutions. Normal Cu level in foliage tissue is 3–20 mg/kg (Table 3.17) and the toxic level in plants is > 25 mg/kg dry matter. Severe Cu toxicity may occur when there is excess Cu accumulation in the roots, but translocation to the top parts of plants is limited. The toxicity of Cu is greater than that of Zn. Animals are susceptible to Cu toxicity, the order of sensitivity being sheep > cattle > swine > poultry (Table 3.17). Molybdenum deficiency may result in Cu toxicity and suitable application of Mo can prevent this type of hazard. Copper toxicity may occur in plants treated with manure produced by animals raised on a high-Cu diet. Monogastric animals and humans are not highly susceptible to Cu toxocity and high Cu levels (such as 200 mg/kg or more) can be ingested when the concentration of Zn and Fe is sufficient.

The recommended upper limit of Cu in animal drinking water is 0.5 mg/l. The acceptable levels in water used for short- and long-term irrigation are 5 and 0.2 mg/l, respectively.

Manganese (Mn). Manganese is required by plants and about 0.5 mg/l is added in nutrient solutions. The content of Mn in soils is 100–4000 mg/kg, with 850 mg/kg being the typical level, and in soil solution it is 0.1–10 mg/l. The range of Mn in plant tissue is 15–150 mg/kg and phytotoxic levels in plant foliage range between 400–2000 mg/kg dry matter (Table 3.17). The recommended upper level for long-term irrigation is 0.2 mg/l (Ayers and Westcot 1985). In acid soils, Mn may become toxic to certain plant species. Manganese toxicity is closely related to iron deficiency. The range and the typical median of Mn levels in digested sewage sludge are given in Table 3.16.

Molybdenum (Mo). Molybdenum is required by plants and is added at low concentrations (about 0.03 mg/l) to nutrient solutions. The normal concentration in soils is 0.2–5 mg/kg, a typical value being 2 mg/kg. The normal concentration of molybdenum in plant tissue is 1–100 mg/kg (Page et al. 1981) and in forage foliage it is 0.1–3 mg/kg. The recommended content in irrigation water is 0.01 and 0.05 mg/l for long- and short-term irrigation, respectively.

Despite the low levels needed by crops, high concentrations of Mo are not toxic to plants; toxicity in forage crops occurs at a level of 100 mg/kg dry matter (Table 3.17). However, Mo may pose a hazard to animals if present in excessive levels in plants. Animal response to Mo intake depends on the species and the age of the individual. Other influencing factors are the concentration of Cu in the food and sulfate level in the diet. Cattle are most susceptible and chicken are least sensitive to Mo toxicity. Molybdenosis in animals, having the same symptoms as Cu deficiency, may develop when forage has Mo levels as low as 5 mg/kg (Page et al. 1981). Since excessive Mo levels leads to copper and phosphorus deficiency, the addition of Cu and P cures this molybdenosis. A combination of high Mo with low Cu levels in forage crops may lead to serious problems. Molybdenum is more available in alkaline soils. Molybdenum derived from organic residues poses no hazard to the food chain in drained soils having a pH lower than 6.

Nickel (*Ni*). Nickel is not required by plants but is needed by animals. The level of Ni in soil is 10–10000 mg/kg, with 40 mg/kg being the typical value. The concentration in the soil solution is estimated to be 0.05 mg/l. Nickel is toxic to some plants at levels between 0.5 and 1.0 mg/l. The normal content of Ni in plant material is 1–10 mg/kg dry matter (Page et al. 1981) and 0.1 and 5 mg/kg dry foliage (Table 3.17). The level of Ni in soils is controlled by the level of Mn or Fe hydrous oxides that reduce the level of soluble Ni in the soil. The level of available Ni is also affected by the presence of organic matter. The range of concentrations of Ni in sludge is wide (2–5300 mg/kg, with a median value of 80 mg/kg). Nickel is toxic to plants at concentrations greater than 50 mg/kg dry matter. Nickel toxicity is acute only in acid soils (Page et al. 1981) and it is reduced by liming. The upper acceptable limits of Ni in irrigation water are 0.2 and 2.0 mg/l for long term and short-term use, respectively.

Lead (*Pb*). Lead is not required by plants and animals, but is potentially highly toxic to animals and less so to plants. The high Pb toxicity is mainly associated with industrial pollution sources rather than with the development of excessive Pb in soils. The level of Pb in soils ranges between 2 and 200 mg/kg, with a typical value of 10 mg/kg, and its concentration in the soil solution is estimated as 0.001 mg/l. The recommended upper level of Pb in animal drinking water is 0.1 mg/l, but problems may occur with levels as low as 0.05 mg/l. The maximum acceptable concentrations in the irrigation water are 5 and 20 mg/l for long- and short-term applications, respectively. The typical median content of Pb in sludge is 250 mg/kg.

Lead reacts with many soil constituents, such as clay, phosphate, carbonate, hydroxides, sesquioxides and organic matter, resulting in low Pb availability. Lead is attenuated by many soils. Jurinak and Santillan-Medrano (1974) attributed the high Pb retention to hydroxides and hydroxy phosphate in acid soils, and to carbonates in calcareous soils. Increasing the soil pH, as well as the CEC and P levels, results in a decrease in soil Pb availability. The lead level in soils is reduced by liming. Large applications of Pb to a soil having a pH level of 5.0 does not result in yield reduction or accumulation of lead in plant tissue. The highest level of Pb is found in plant roots, and its value in other plant parts diminishes in the order: shoot, fruit and seeds. Its typical concentration in crop tissue is 0.1–10 mg/kg (Page et al. 1981) and 2–5 mg/kg in forage (Table 3.17). Under normal conditions of sludge application, when a large quantity of P is also added and at soil pH above 5.5, the movement of Pb to plant tops and seeds is nil.

Selenium (Se). Selenium is not required by plants, but it is essential for animals at low concentrations (0.04 mg/l). It becomes toxic to animals at levels of 4–5 mg/l. The range of Se in soils is 0.1–2.0 mg/kg with 0.5 mg/kg being typical; the Se level in plant tissue is 0.02–2.0 mg/kg with toxic levels above 50 mg/kg. The maximum recommended content in animal drinking water is 0.05 mg/l, and 0.02 and 0.05 mg/l in long-term and short-term irrigation, respectively. The level of Se in soil solution is 0.001–0.01 mg/l and that in digested sludge ranges from about 2 to about 17 mg/kg.

The solubility of Se in soils is least under acid conditions. Under neutral and alkaline conditions selenate ion is formed, which does not form insoluble salts and stable sorption complexes with soil clay. The toxic effect of selenium on animals was detected in dry regions, the soils of which had developed from parent materials containing selenium. Selenium pollution of drainage water became a severe problem in certain areas in California, but the source of Se in this case were natural rock and soil parent materials (Oster et al. 1989). The level of Se in wastewater is usually very low.

Vanadium (V). Vanadium is required by certain algae. It is not required by plants, but is toxic to many plants at a relatively low concentration. Its level in soils ranges from 20 to 500 mg/kg with a typical concentration of 100 mg/kg (Page et al. 1981). The tolerable concentration in animal drinking water and recommended maximum concentration in long-term irrigation water is 0.1 mg/l. Its levels in the tissue of forage crops is 0.1–1 mg/kg, and it becomes toxic at a concentration of 10 mg/kg (Table 3.17).

Zinc (Zn). Zinc is essential for plants and animals, and is added at the rate of about 0.5 mg/l to nutrient solutions. Its concentration in soils is 10–300 mg/kg, with 50 mg/kg being the typical level (Page et al. 1981). Zinc concentration in soil solution is <0.005 mg/l. The typical range of Zn in plant tissues is 15–200 mg/kg. The recommended maximum value for drinking water for livestock is 24 mg/l, and for water used for long-term and short-term irrigation 2 and 10 mg/l, respectively. Zinc is toxic to plants at different concentrations, typical values being >200 mg/kg. The toxic levels in forage foliage is 500–1500 mg/kg (Table 3.17). Zinc toxicity is a greater problem for plants than for animals or humans.

The prevalent zinc form in the solution of acid soils in Zn^{2+}; this is the Zn form taken up by plants. The main factors responsible for Zn retention in soils are sorption by clay, hydrous oxide surface reactions, and chelation by organic matter. Zinc availability is reduced by liming more than that of Cd, Cu or Ni. The difference between normal and toxic levels in plant tissue is wide enough to create a safety margin. Sheep and cattle are more sensitive than other domestic animals to Zn toxicity. At pH greater than 6, sludge application will not cause problems unless exceptionally high levels of Zn are added.

3.7 Organic Compounds

3.7.1 Organic C and N

Standard treatment methods remove most of the organic matter from raw sewage. Consequently, the typical COD, BOD, or TOC levels found in effluents (see Sect. 2.8) are much lower than in the original sewage, and the quantities of organic C added to soils through effluent irrigation are also relatively low. For instance, the amount of organic C applied through an annual addition of 1000 mm of typical sewage effluent containing a total organic C level of 50 mg/l is only 500 kg/ha.

Since the amount of N in sewage effluents is high, the C/N ratio for the organic components is low (typical values are 5), and a release of mineral N from organic substances is favored. The organic C/total N ratio is much lower than the organic C/organic N ratio. Application on sewage effluent having a high BOD value reduces the oxygen level in the soil, and consequently, denitrification may be enhanced (see Sect. 3.3).

Soil organic matter affects soil structure and its stability, and is a source for nutrient elements essential for plant growth. The C cycle in soil consists of manifold reactions in which numerous microorganisms take place (Alexander 1977). Many reviews deal with C transformations in the soil-plant system, some of which are specific to soils treated with organic waste (e.g., Gilmour et al. 1977; Stevenson 1985).

3.7.2 Trace Organics

Recent developments in laboratory equipment and analytical methods have enabled the identification of organic substances occurring in soil and water in extremely low amounts (Grigoropoulos and Smith 1971; Cooper 1981; Chang and Page 1985). On the other hand, certain trace organics present in soil and water are known to be toxic. These may have entered the soil through the use of chemicals, such as pesticides and herbicides. Alexander (1981) discussed the microbial synthesis of hazardous substances, which have been produced from precursors in aqueous sediments or polluted soils. The formation of certain carcinogenic material (methylmercury, dimethylarsine, dimethylselenide, and nitrosamine) was recently discussed by Stevenson (1985). Due to the great health concern caused by trace organics, much analytical and experimental work has been carried out to obtain reliable information on the presence of trace organics in soil and water. Much of the relevant research involves the behavior of agrochemicals and industrial products in soils and their movement into aquifers (Rao and Jessup 1983).

Being a product of various domestic and industrial activities, sewage water contains numerous synthetic organic compounds. Although much of this organic material is removed through standard treatment methods, organic residues are still identified in variously treated effluents (see Table 2.9).

The great concern for possible health hazards resulting form the presence of trace organics in the water is reflected in the relevant publications discussing various aspects of sewage effluent reuse. Chang and Page (1985) concluded that the risks from trace organics associated with the use of sewage effluent for irrigation are not greater than those presented by other sources of water. This is due to the reduction in the number of substances present in the effluent and in their concentration following treatment. Moreover, various soil reactions, attenuation processes, similar to those occurring in pesticide-treated soils, further reduce the level of trace organics in the soil-water system. Finally, the quantities of trace organics added through sewage effluents are usually smaller than those applied by means of standard pesticide treatment.

3.7.3 Reactions of Trace Organics in Soil

Trace organics derived from different sources, including sewage effluent, undergo adsorption, volatilization and biodegradation reactions in the soil, which greatly reduce the level of the original material in the effluent. Much of the available information on this subject was obtained from pesticide-treated soils (Rao and Jessup 1983; Jury et al. 1983, 1984a–c; Chang and Page 1985).

The adsorption of pesticides on soil materials is described by the Freundlich and Langmuir equations (see Sect. 3.4). These equations give the distribution of trace organics between the adsorbed and solute components. Under equilibrium conditions, the sorption of toxic organic substances (TOS) on soils and sediments is described by the Freundlich equation (3.30) (Rao and Jessup 1983). The sorption coefficients are empirical constants, usually obtained by curve fitting the equations to measured data. The value of the partition or distribution coefficient K is a measure of pesticide or TOS sorption on the soil. For nonionic and polar pesticides, the soil organic content (% OC) is the best indication of the level of adsorption. For ionic or ionizable substances the effect on sorption of pH is considerable.

Volatilization of trace organics from dilute aqueous solutions is determined by the partitioning coefficient (K_w) between water and air (Chang and Page 1985). K_w is defined as the ratio of the equilibrium concentration in the water (μg/ml) to the corresponding concentration in the air (μg/ml) at 25 °C. The lower the K_w, the greater is the volatilization. Since it is difficult to achieve steady-state conditions, K_w in the air is determined experimentally. Estimates are made when necessary data are unavailable. Page and Chang (1985) present data on soil adsorption constants (K_{OC}), water-air partitioning coefficient (K_w) and other valuable information on the behavior of different trace organics and pesticides in the soil.

Biodegradation is responsible for reduction in the level of trace organics in the soil. The type of microorganisms, soil properties and the chemical properties of the organic substances involved determine the rate of decomposition. At low substrate concentrations, degradation of organic substances follows a first-order reaction. The presence of available organic C derived from other sources enhances the decomposition of trace organics in the soil, while adsorption reduces its rate.

Chang and Page (1985) reviewed evidence of the occurrence of bioaccumulation, i.e., uptake of trace organics through plant roots. Although such uptake has been detected, the available information indicates that this process does not play an important role in soil-plant systems.

3.8 Suspended Solids

Suspended solids in secondary effluents are mainly of biological origin. Their concentration and composition depend on the level of treatment and particularly on the degree of separation of sludge from the treated sewage in the final sedimentation process. Turbidity in excess of 20–30 mg/l (a common value for biological treatments) may be found in over/or underloaded systems. Suspended particles in secondary effluent leaving the treatment plant contain various microorganisms, such as bacteria, protozoa, rotatoria and phytoflagellata, ranging in size from colloidal to several hundred microns in size. Organic and mineralized flocs of 1 mm or even larger in size (usually of porous structure), as well as clay particles and other inorganic colloidal matter, are also present.

In general, suspended matter in secondary sewage effluent used for irrigation does not have any adverse affect on soil, plant or environment. In some cases, though, with effluents of high organic and nutrient content used for surface flood irrigation, development of microbial growth on the soil surface may reduce the soil infiltration rate and cause aeration problems. Addition of silt particles, due to either erosion of earth canals or improper pumping from earth strogate reservoirs, may cause clogging of soil pores, form a crust and affect infiltration and aeration. Friction of such rigid particles suspended in the irrigation water may interfere with proper functioning of regulating, control and measuring devices in the irrigation systems. The most important hazard associated with suspended matter is the clogging of small openings and/or orifices in the irrigation system. Particularly vulnerable to clogging are drip irrigation systems (see Table 2.10). In most cases, filtration and chemical treatment, such as chlorination, of the irrigation water just before use may help overcome these problems. Further discussion on clogging of drip irrigation systems and filtration can be found in Chapt. 6.

It is noteworthy, however, that in most cases secondary effluent intended for irrigation is stored in reservoirs for varying periods and in some cases is mixed with water from various sources. During storage, different kinds of algae and aquatic organisms generally develop according to specific conditions in the reservoir. The diversity of the biological population makes it difficult to generalize on the effect of clogging on irrigation systems, and on the means to overcome clogging hazards.

3.9 Health Aspects

3.9.1 Sources of Health Problems

There are two main components of sewage effluents which are related to public health, i.e., pathogenic organisms and hazardous chemicals (Kowal 1983; Fig. 3.24). Environmental and food chain aspects of trace elements were discussed in Sect. 3.6.5. *Pathogenic organisms* (viruses, bacteria, protozoa and helminth eggs, see Sect. 2.8) may be associated with the transmission of disease to farm workers and other people and to livestock exposed to the effluent either by incidental physical contact, by inhalation of aerosols from spray-irrigated fields or by consuming crops irrigated with effluent. *Hazardous chemicals* may reach groundwater or surface water or contaminate crops (Kristensen and Bond 1977; Reed 1982). It is important, therefore to define criteria and formulate regulations for using sewage effluent for irrigation (WHO 1973; Bouwer 1982; Ayers and Westcot 1985; Pescod 1986; Shuval 1987).

3.9.2 Health Criteria

The current quality criteria (WHO 1973) aim at protecting public health against diseases that could possibly be derived from raw and partially treated sewage water. In the absence of well-established epidemiological information, the health criteria being in use today are based on assumptions regarding the possible effects of sewage effluent, rather than on well-documented evidence of actual problems caused by irrigation with sewage effluent. Recent information derived from epidemiological surveys resulted in a tendency to ease the restriction imposed by

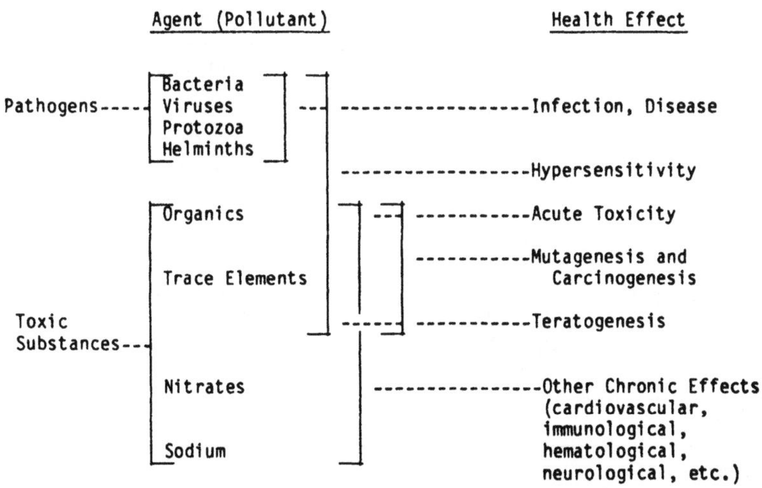

Fig. 3.24 Major health effect of pollutants (Kowal 1983)

the current stringent standards (Shuval 1987). More details on this development in quality guidelines are given below (Sect. 3.9.3).

The current quality criteria are based on the degree of risk of infection by sewage-borne pathogens. This risk depends on many factors, including the efficiency of removal or inactivation of the disease causing organisms by the treatment, their survival in the effluent, the soil and the plants, and the degree of infection considered to cause disease in humans (Shuval 1977).

The greatest risk of infection or illness is through effluent ingestion. Such a situation may occur through inadvertent drinking from the supply systems. Therefore, all pipes and appurtenances used for irrigation with reclaimed effluent should be color-coded and all the valves and outlets should be tagged with appropriate warnings.

Diseases may be transmitted to humans by handling or consumption of contaminated crops, especially where irrigated with raw sewage. Important factors in this regard are the persistence of pathogens on crops at the time they are consumed by humans or animals and the persistence of pathogens in the soil and water. Under favorable conditions, pathogens may survive for long periods. The survival of organisms depends on many factors, such as type and number of pathogens, as well as temperature, humidity, sunlight intensity and competition from other microorganisms. Relevant papers (Bryan 1977; Frankenberger 1985; Pescod 1986, and Table 3.18) have reported survival for extended periods of enteric pathogens on vegetables (weeks) and moist soils (months and even years). Larkin et al. (1976) demonstrated the persistence of polio virus on lettuce and radish for up to 36 days. In a study carried out in Israel, Shuval (1978) described

Table 3.18 Survival of pathogens in soils[a] and crops[b]

Organism	Survival time in soils (days)	Survival on crops (days)
Coliforms	38	<30 (<15)[c]
Streptococci	35–63	
Fecal streptococci	26–77	
Salmonellae	15–>280	<30 (<15)
Shigella spp.	—	10 (<5)
Vibrio cholerae	—	5 (<2)
Salmonella typhi	1–120	
Tubercle bacilli	>180	
Leptospira	15–43	
Entamoeba histolytica cysts	6–8	<10 (<2)
Enteroviruses	8–175	<60
Ascaris ova	Up to 7 years	
Hookworm larvae	42	<60 (<30)
Brucella abortus	30–125	
Q-fever organisms	148	

[a]After Frankenberger (1985).
[b]After Pescod and Alka (1984).
[c]The usual survival times are shown in parentheses.

internal contamination of plant tissues with viruses transported to the leaves via roots. However, these results were obtained under extreme contamination ocnditions, where damaged roots were exposed to an extremely high concentration of the virus in the soil. Contamination was not detected in tissue of undamaged plants not exposed to a high virus population.

Sunlight exerts a lethal effect on all microorganisms involved in soil surface or plant contamination. More than 99% of the detectable viruses and bacteria were eliminated after 2 days of exposure to sunlight. Such findings support the regulation that a suitable time interval be maintained between effluent irrigation and grazing or crop handling (Sepp 1971).

Pathogens retained in the soil surface layer may be transported in runoff water, may percolate into groundwater or may be carried by insects, rodents and birds. Movement of bacteria and viruses in the soil was discussed by Frankenberger (1985) and data are shown in Tables 3.19 and 3.20.

Filtration, adsorption and sedimentation processes taking place in soils make its matrix an efficient medium for eliminating bacteria, helminth eggs and protozoan cysts. Relevant studies show that bacteria and, to an even greater extent, viruses are confined to the top few centimeters of the soil surface layer and never reach groundwater (Gerba et al. 1975). At low percolation rates, which are normal in irrigated agricultural soils, pathogen immobilization and inactivation is very efficient. However, from a public health standpoint, provision for adequate and reliable disinfection has been considered as an essential feature of treatment processes of sewage effluent intended for agricultural use (Shuval 1975). Chlorination is the most commonly used disinfection method for this purpose

Table 3.19 Movement of bacteria through soils in relation to wastewater application (After Frankenberger 1985)

Nature of fluid	Organisms	Media	Maximum distance traveled (m)	Time of travel
Tertiary treated wastewater	Coliforms	Fine to medium sand	6.1	—
Secondary sewage effluent on percolation beds	Fecal coliforms	Fine loamy sand to gravel	9.1	—
Primary sewage in infiltration beds	Fecal *Streptococci*	Silty sand and gravel	183	—
Inoculated water and diluted sewage injected subsurface	*Bacillus stearothermophilis*	Crystalline bedrock	28.7	24–30 h
Sewage in buried latrine intersecting groundwater	*Bacillus coli*	Sandy and sandy clay	10.7	8 weeks
Canal water in infiltration basins	*Escherichia coli*	Sand dunes	3.1	—

Table 3.20 Movement of viruses through soil in relation to wastewater application (After Frankenberger 1985)

Virus type	Nature of fluid	Nature of medium	Flow rate	Distance of travel	Percentage of removal
T1, T2, f2	Distilled water with added salts	9 types of soils from California	0.078–0.313 ml/min	45–50 cm	> 99
Poliovirus 1	Distilled water, 10^{-5} N Ca and Mg salts	Dune sand	1–2 ml/min	20 cm	99.8–99.9
Poliovirus 2	Distilled water	Low humic latosol	4.1–5.7 $m^3 m^{-2} day^{-1}$	4–15 cm	96–99.3
Poliovirus 2	Secondary effluent	Sandy gravel	—	60 m	100
Coxsackie	Spring water	Garden soils	—	90 cm	50
T4	Distilled water	Low humic latosol	4.1–5.7 $m^3 m^{-2} day^{-1}$	4–15 cm	100
T7	Secondary treated	Sandy forest	—	19.5 cm	99.6
Indigenous enteric	Secondary effluent	Loamy sand soil	Intermittent avg: 0.02 cm/min	3–9 m	100

and can also be applied to partially-treated effluent and is easily incorporated into most irrigation systems (Engelbrecht 1980; see also Sect. 6.4).

The use of sprinkler or spray irrigation results in wind drift of aerosols made of small effluent water droplets carrying pathogenic bacteria and viruses a considerable distance from the irrigated site (Bausum et al. 1983). For instance, a recent publication of Camann et al. (1988) reported a significant increase in the concentration of fecal coliforms, fecal streptoccoci, mycobacterium and coliphage in the air, at least 200 m downwind from the application site. The effluent used was a low quality wastewater taken directly from the treatment plant. Enteroviruses at a higher levels than previously reported in the US or Israel were detected 44–60 m downwind. Storage in reservoir reduced the relevant concentration of indicator organisms in the air by two orders of magnitude. It is therefore recommended that appropriate buffer zones (generally 100–500 m; Young 1980) be established between spray-irrigated fields and residential areas or other sites involved in intensive public activity. The best protection against aerosols hazards is the use of drip or surface irrigation. Drip irrigation and several surface irrigation systems (e.g., leveled basin) can be operated automatically, thus minimizing direct human contact with the effluent. In most surface irrigation systems, there is a risk of direct contamination of farm workers, resulting primarily from improper personal hygiene and walking barefoot on wet ground. It is therefore recommended that people entering effluent-irrigated areas always wear shoes. Recent developments in subsurface drip irrigation (Oron et al. 1988) make this method very attractive for effluent application, since it greatly reduces direct contact with effluent or contamination of crop yield.

3.9.3 Health Guidelines for the Use of Sewage Effluent in Agriculture

The discovery of the association between sewage irrigation and health problems resulted in the development of standards and guidelines aimed at protecting public health against sewage-borne diseases. In 1918 the first relevant publication of health regulations appeared in California, and later in other countries (Shuval 1987). The present California regulations for crop irrigation (Table 3.21) demand a high degree of effluent treatment, 23 and 2.2 coliforms MPN (Most Probable Number/100 ml), for processed products and crops eaten raw, respectively. The latter coliform level is "more or less parallel to that required for drinking water" (Shuval 1987). Such effluent requires a complete biological treatment followed by a heavy chemical disinfection. A less stringent guideline for effluent irrigation has been issued by the WHO (1973); for instance, the demand for crops eaten raw is 100 coliforms/100 ml in 80% of the samples taken for laboratory analysis (Table 3.22). Similar standards are in use in different states or countries (Bower 1982).

The Israel Ministry of Health (1979) permits the use for irrigation of secondary effluent containing at least 0.5 mg oxygen/l, free of toxic compounds hazardous to human health either due to direct contact with the effluent or to contact with the crop in the field or after harvest. The following crops may be irrigated:

1. Nonedible crops (e.g., cotton, seed crops and fodder)
2. Industrial crops (e.g., grains and sugar beet), but not crops intended for canning
3. Groves and forests not used for recreation or camping
4. Dates with under-tree irrigation, provided there is no gathering of fallen fruits

Table 3.21 Wastewater treatment and quality criteria for irrigation (California) (Crook 1985)

Treatment level	Coliform limits	Type of use
Primary		Surface irrigation of orchards and vineyards, fodder, fiber and seed crops
Oxidation and disinfection	≤ 23/100 ml	Pasture for milking animals Landscape impoundments Landscape irrigation (golf courses, cemeteries, etc.)
	≤ 2.2/100 ml	Surface irrigation of food crops (no contact between water and edible portion of crop)
Oxidation, coagulation, clarification, filtration[a], and disinfection	≤ 2.2/100 ml max. = 23/100 ml	Spray irrigation of food crops
		Landscape irrigation (parks, playgrounds, etc.)

[a] The turbidity of filtered effluent cannot exceed an average of 2 turbidity units during any 24-h period.

Table 3.22 Suggested treatment processes to meet the given health criteria for wastewater reuse[a]

Wastewater use	Irrigation			Recreation		Industrial reuse	Municipal reuse	
Health criteria[b]	Crops not for direct human consumption (A + F)	Crops eaten cooked: fish culture (B + F or D + F)	Crops eaten raw (D + F)	No contact (B)	Contact (D + G)	(C or D)	Non-potable (C)	Potable (E)
Primary treatment	3	3	3	3	3	3	3	3
Secondary treatment		3	3	3	3	3	3	3
Sand filtration or equivalent polishing methods		1	1		3	1	3	2
Nitrification						1		3
Denitrification								2
Chemical clarification						1		2
Carbon adsorption								2
Ion exchange or other means of removing ions						1		2
Disinfection[c]		1	3	1	3	1	3	3[c]

[a] After WHO (1973).
[b] Health criteria: A, freedom from gross solids, plus significant removal of parasite eggs; B as A, plus significant removal of bacteria; C as A, plus more effective removal of bacteria, plus some removal of viruses; D, not more than 100 coliform organisms per 100 ml in 80% of samples; E, no fecal coliform ortanisms in 100 ml, plus no virus particles in 1000 ml, plus no toxic effects on man, and other drinking-water criteria; F, no chemicals that lead to undesirable residues in crops or fish; G, no chemicals that lead to irritation of mucous membranes and skin. In order to meet the given health criteria, processes marked 3 will be essential. In addition, one or more processes marked 2 will also be essential, and further processes marked 1 may sometimes be required.
[c] Free chlorine after 1 h.

5. Nuts and almonds with under-canopy irrigation, provided picking is at least 2 weeks after the last irrigation
6. Deciduous fruits (apples), which are drip-irrigated, provided there is no gathering of fallen fruit
7. Banana, which is drip irrigated

The following health standards have been issued in 1983 by the health authorities of Arizona (Bouwer and Idelovitch 1987): the geometric mean for fecal coliforms (CFU/100 ml; 5 samples minimum) is 1000 for orchards, fiber, seed and forage crops, pastures, livestock watering, and processed food crops; 200/100 ml for landscape of restricted access; 25/100 ml for landscape areas of open access, and 2.2/100 ml for crops consumed raw. The number of enteric virus (PFU/40 l) is 125 and 1 for landscape areas of open access and crops eaten raw, respectively. A nondetectable (ND) level is the standard for *Entamoeba hystolytica* in effluent used to irrigate crops eaten raw, and for *Ascaris lumbricoides* (roundworm eggs) in landscape areas of open access and crops consumed raw. An ND level is also required for common large tapeworm in pasture and livestock watering. The effluent turbidity allowed for landscape areas of open access and crops eaten raw is 5 and 1 nephlometer turbidity units (NTU), respectively.

Due to their effect on the cost of effluent treatment, health regulations for effluent irrigation were often "a compromise between what is theoretically desirable for public health and what is practically achievable" (Bouwer and Idelovitch 1987); too stringent regulations can increase the price of the treated effluent to a level that makes effluent irrigation not practical. Shuval et al. (1986) carried out an extensive survey in many developing countries to determine if the current health standards can be justified by epidemiologial facts and whether there is a reason to alter them. The results of this study indicated (Shuval et al. 1986; see also Shuval 1987) that the previous restrictions imposed on effluent use for the protection of public health were overly conservative.

Health problems resulting from wastewater use were limited and almost exclusively associated with the consumption of fresh produce irrigated with untreated sewage water. Furthermore, irrigation with partially treated wastewater, even though not approaching the quality of drinking water, was not associated with negative health effects. Previously published articles have shown no difference in the occurrence of intestinal disease among members of Israeli Kibbutz settlements using partially-treated sewage effluent for irrigation, compared with those using fresh irrigation water (Shuval et al. 1986). Most of the epidemiological studies conducted on residents of communities subjected to sewage-effluent aerosols have not detected any correlation between exposure to aerosols and contraction of diseases (Pahren and Jakubowski 1980).

Smith (1982, cited in Strom 1984) reported results of a 3-year study in which activated-sludge treated sewage effluent (obtained from a plant near Melbourne, Australia) was used to irrigate 7 vegetable crops. The effluent, the vegetables and several soil samples were tested for bacteria, viruses and heavy metals. No difference in the bacteriological contamination between effluent and fresh water irrigation was detected. Viruses could not be detected in soil surface and subsurface layers, respectively, 1 and 2 days after irrigation with the reclaimed effluent. The quantities of bacteria detected on vegetables irrigated with the reclaimed water were not significantly different from those found on noneffluent-irrigated vegetables in the vicinity of Melbourne. It was, therefore concluded that, under the conditions of the trial, there was no risk of viral infection from the consumption of vegetables irrigated with reclaimed effluent.

The available epidemiological information was reevaluated by a group of environmental experts and epidermiologists, meeting in Engelberg, Switzerland in 1985 (the Engelberg Report 1985). The group accepted the main findings of Shuval et al. (1986) and the analysis of Blum and Feechem (1985) of the epidemiological health aspects of using human waste and sludge in agriculture and aquaculture. It concluded that "current guidelines and standards for human waste use are overly conservative and unduly restrict project development, thereby encouraging unregulated human waste use".

The tentative microbiological quality guidelines suggested in the Engelberg Report (1985) is given in Table 3.23. The level of fecal coliforms allowed for edible crops is less than 1000/100 ml, and is not applicable to nonedible crops. The number of intestinal nematodes acceptable in both restricted and unrestricted irrigation is less than 1 viable egg/l. Shuval et al. (1986) have shown (Fig. 2.4) that such effluent quality can be obtained after a few days or weeks of detention in stabilization ponds. The advantage of such techniques is the availability of low cost, efficient method for sewage treatment that is suitable for developing countries.

A recent report (WHO Scientific Group 1989) presents the collective view of an international group of experts and gives (Table 3.24) microbiological quality

Table 3.23 Tentative microbiological quality guidelines for treated wastewater reuse in agricultural irrigation[a] (Engelberg Report 1985; Shuval 1987)

Reuse process	Intestinal nematodes[b] (geometric mean no. of viable eggs per liter)	Fecal coliforms (geometric mean No. per 100 ml)
Restricted irrigation[c]		
Irrigation of trees, industrial crops, fodder crops, fruit trees[d] and pasture[e]	< 1	not applicable
Unrestricted irrigation		
Irrigation of edible crops, sports fields and public parks[f]	< 1	< 1000[g]

[a] In specific cases, local epidemiological, sociocultural, and hydrogeological factors should be taken into account, and these guidelines modified accordingly.
[b] *Ascaris*, *Trichuris* and hookworms.
[c] Minimum treatment equivalent to at least a 1-day anaerobic pond followed by a 5-day facultative pond or its equivalent is required in all cases.
[d] Irrigation should cease 2 weeks before fruit is picked, and no fruit should be picked off the ground.
[e] Irrigation should cease 2 weeks before animals are allowed to graze.
[f] Local epidemiological factors may require a more stringent standard for public lawns, especially hotel lawns in tourist areas.
[g] When edible crops are always consumed well-cooked, this recommendation may be less stringent.

Table 3.24 Recommended microbiological quality guidelines for wastewater use in agriculture[a] (WHO Scientific Group 1989)

Category	Reuse conditions	Exposed group	Intestinal nematodes[b] (arithmatic mean No. of eggs/l[c])	Faecal coliforms (geometric mean No. per 100 ml[c])	Wastewater treatment expected to achieve the required microbiological quality
A	Irrigation of crops likely to be eaten uncooked, sports fields, public parks[d]	Workers, consumers, public	≤ 1	≤ 1000[d]	A series of stabilization ponds designed to achieve the microbiological quality indicated or, equivalent treatment
B	Irrigation of cereal crops, industrial crops, fodder crops, pasture and trees[e]	Workers	≤ 1	No standard recommended	Retention in stabilization ponds for 8–10 days or equivalent helminth and faecal coliform removal
C	Localized irrigation of crops in category B if exposure of workers and the public does not occur	None	Not applicable	Not applicable	Pretreatment as required by the irrigation technology, but not less than primary sedimentation

[a] In specific cases, local epidemiological, sociocultural and environmental factors should be taken into account, and the guidelines modified accordingly.
[b] *Ascaris* and *Trichuris* species and hookworms.
[c] During the irrigation period.
[d] A more stringent guideline (≤ 200 faecal coliforms per 100 ml) is appropriate for public lawns, such as hotel lawns, with which the public may come into direct contact.
[e] In the case of fruit trees, irrigation should cease 2 weeks before fruit is picked, and no fruit should be picked off the ground. Sprinkler irrigation should not be used.

guideline for wastewater use in agriculture. The authors stress the fact that the suggested guidelines are based on firm epidemiological evidence rather than potential health risk. The less stringent regulations suggested by the group allow the use of relatively simple treatment techniques that are better applicable to developing countries. Other reports support the view presented by report of the WHO Scientific Group (1989). Where, due to economic constraints, advanced purification of effluent followed by unrestricted irrigation is not achievable today, partial treatment for restricted irrigation is recommended. For example, Pescod (1986) concluded that "Generally it is not advisable to rely on sustained high level of performance from advanced technological systems in developing countries and in irrigation reuse of sewage this can be dangerous"; and furthermore "unrestricted irrigation with highly-treated effluent is unlikely to be consistently safe in developing countries and will not normally produce the most appropriate or economic reuse scheme". Using his experience in Saudi Arabia, Ali (1987) also suggested a middle-of-the-road approach for the use of sewage effluent for crop irrigation in arid regions.

Restricted vs Unrestricted Irrigation. Bouwer and Idelovitch (1987) define two types of effluent water use, restricted and unrestricted. Restricted irrigation refers to the use of low-quality effluent under specific agricultural conditions. The major aspects involved in restricted irrigation are soil properties, location characteristics such as the presence of aquifers, irrigation, fertilization and harvesting methods, distance from roads, residential areas and pipelines carrying potable water. Bouwer and Idelovitch (1987) concluded that, although restricted irrigation has the advantage of simplicity and low cost treatment, it is generally applicable to small quantities of effluent, and specific locations where the irrigated crops are well defined and not likely changed.

The authors recommend not to use low quality water on extensive areas, since problems greatly increase as the size of the irrigated land increases. Where such problems can occur, a high-quality effluent and unrestricted irrigation are preferred; this gives the possibility of using effluent safely for any crop in any area and also to allow livestock watering. Using such waters, direct contact and even accidental drinking are not likely to cause problems. This policy is followed in Israel, where the effluent used for irrigation of large areas in the southern part of the country originating from the Tel Aviv metropolitan district is given a high level of treatment (Idelovitch 1978).

Chapter 4 Practical Use of Effluent Irrigation

4.1 Introduction

Irrigation with sewage effluent has been successfully practiced over a long period (Pound and Crites 1973b; Harlin 1979; Page et al. 1983; Shuval et al. 1986). The earliest documented sewage farms were those of Bunzlau, Germany, and Edinburgh, Scotland, which were in operation in 1531 and around 1650, respectively. Many other farms, such as those in London, Paris (1868), Berlin (1876) and Melbourne (1897), were established later. Sullivan et al. (1973) carried out an extensive survey on land application of sewage effluent of various municipal and industrial origins. According to their findings, land application of wastewater from community and industrial processing sources is widespread in the US and many other countries throughout the world, and under proper conditions, is a workable alternative to advanced or tertiary treatment of municipal wastes. Page et al. (1983) concluded that land treatment systems are now successfully operated throughout the world. Moreover, problems can be controlled through proper engineering operation and management. Some representative examples of waste water use, including long term experiments are presented below.

The information covers a variety of soil, climate and crop conditions to give a well-balanced picture of present-day use of sewage effluent for irrigation, especially where this source of water is economically important.

4.2 Arid and Semi-Arid Regions

Irrigation is standard practice under arid and semi-arid climates, where intensive cropping cannot depend on rainfall alone. In dry regions there is usually a shortage of irrigation water, thus sewage water, as well as other low quality water resources, may become important sources of water and nutrients. Large quantities of sewage effluent are used for irrigation in southern and western USA, Mexico, Australia, dry regions of Africa and Asia, and Israel.

4.2.1 Arizona

Day (1973) has summarized much of the experimental work and the experience gained with secondary effluent in Arizona, mainly at the Arizona Agricultural Experiment Station of the University of Arizona, Tucson. He reported on experiments which compared different combinations of well water plus fertilizer with wastewater alone. A typical set of treatments studied in field experiments reported by Day was (1) well water with no additional fertilizer (control); (2) well water plus recommended amounts of fertilizer (112 kg N and 37 kg P/ha); (3) well water with macronutrient levels equal to those of wastewater (224 kg N, 76 kg P and 93 kg K/ha); and (4) wastewater with no additional fertilizer. Crop yields were greatly affected by these treatments, e.g., that of oats (*Avena sativa* L.) was 1.5 times greater using wastewater than with well water supplemented with recommended amounts of commercial fertilizer (see also Day and Tucker 1959). The nutrient values of the oats forage obtained from the various treatments were similar. Similar yields of barley hay were obtained from plants irrigated either with sewage effluent or with well water plus standard fertilization (Day and Tucker 1960).

Day and Kirkpatrick (1973) found no significant differences in yield and protein content between oat plants irrigated with municipal wastewater *vs* fresh water plus commercial fertilizer in a container experiment (Table 4.1).

The long-term effects of furrow irrigation with municipal wastewater *vs* well water with recommended N, P and K fertilizer levels were studied in a field experiment carried out on a silty loam soil in Arizona (Day 1973). Barley, cotton, lettuce and sorghum were grown in rotation for 14 years. Irrigation with municipal wastewater resulted in lower infiltration rates and a higher modulus of rupture, indicating adverse effects on soil physical properties. The salinity level in the wastewater-treated soil increased, but the levels of N and P in the yield obtained from the different treatments were similar. The authors did not detect any unfavorable effects on the soil that could not be corrected by minor changes in crop culture.

Another field study carried out in Arizona (Day et al. 1981) compared a 1:1 mixture of municipal wastewater and well water with well water alone, using cotton as a test crop. The well water contained 4.6 g soluble salts per l and 22 mg N/l (21 mg of which was NO_3-N). The concentrations of salts, N and P in the effluent were 3.4, 40 (6 was NO_3) and 4 mg/l, respectively. The addition of 56 kg N/ha as fertilizer to the well water increased the total N added with this

Table 4.1 Effect of well water + NPK fertilizer vs wastewater on yield of oat forage and its grain yield and quality (Day and Kirkpatrick 1973)

Irrigation and fertilizer management	Plant height (cm)	Tillers (No./plant)	Green forage	Dry forage	Total protein: Pasture forage	Total protein: Grain	Grain yield (g/m^2)
			(kg/m^2)		(g/kg)		
Well water + NPK[a]	58	11	2.05	0.41a	250	160	18
Wastewater	57	10	2.22	0.34b	270	180	25

[a] Different letters next to figures indicate significant differences at $P < 0.05$ levels.

Table 4.2 Effect of well water amended with fertilizer (see text) and 1:1 mixture of wastewater and well water on some yield indices of cotton (cv Deltapine) and alfalfa

Treatment	*Cotton*[a] *Average for 1974 and 1975 seasons* Plant height	Yield of seed cotton	Yield of lint cotton	Seed weight
	(cm)	(kg/ha)	(kg/ha)	(g/25 bolls)
Well water	91b[c]	4087b	1580b	76.8b
Mixture	119a	4534a	1676a	83.0a
	Alfalfa[b] Plant height	Yield of hay	Total protein content	
	(cm)	(kg/ha)	(g/kg)	
Well water	63b	3672b	205	
Mixture	72a	4778a	215	

[a] Day et al. (1981).
[b] Day et al. (1982).
[c] Different letters next to the figures indicate significant differences at $P < 0.05$ levels.

treatment to 334 kg N/ha, compared with 488 kg N/ha in the well water-wastewater mixture. Both the vegetative growth and the yield of cotton were greater for the mixture than using well water alone (Table 4.2), while no significant difference in the quality of lint was found. In this experiment the addition of wastewater to the well water diluted the total salt content in the water. Based on these results, Day et al. concluded that wastewater can be used effectively as a source of irrigation water and plant nutrients in the commercial production of cotton in Arizona and, possibly, in similar environments throughout the world.

Higher yields of flood-irrigated alfalfa (*Medicago sativa* L.) were obtained with 1500 mm of a 1:1 mixture of municipal wastewater with well water than well water amended with 100 kg P/ha (Table 4.2; Day et al. 1982).

Sorghum was another important crop tested within the framework of the Arizona studies (Day and Tucker 1977). The yield and some quality indices obtained from container grown plants subjected to various irrigation treatments are presented in Table 4.3. The vegetative development and grain yield were greater using wastewater, which added to the soil 224, 73 and 140 kg/ha, of N, P and K respectively. Well water plus fertilizer, supplying the same amounts of N, P and K as the wastewater, resulted in shorter plants and lower grain yield. Well water plus the recommended fertilizer level (112, 35 and 1 kg N, P and K/ha, respectively) resulted in the lowest grain yield.

The Buckeye Irrigation Company uses about 49 million m^3 of secondary effluent originating in Phoenix and 62 million m^3 of groundwater per year to irrigate about 7200 ha of cropland (Sabol et al. 1987). Since the local well water contains more salt (3000–5000 mg/l) than the effluent (800–1000 mg/l), the

Table 4.3 Effect of well water amended with fertilizer (see text) and treated municipal wastewater on some yield indices of sorghum (Day and Tucker 1977)

Treatment	Plant height (cm)	Grain yield (g/m^2)	Weight of 1000 seeds (g)	Total protein (g/kg)
Well water plus recommended NPK fertilizer[a]	82b	375c	21a	116a
Well water plus NPK equal to levels in wastewater	81b	550b	21a	115a
Wastewater	86a	725a	22a	100a

[a] Different letters next to the figures indicate significant difference at $P < 0.05$ levels.

dilution of the saline water with effluent decreases the load of salt per area unit in the region. On the other hand, the level of NO_3-N in the groundwater is gradually increasing, probably due to fertilizer application and the use of sewage effluent containing 20–30 mg N/l.

4.2.2 California

Irrigation with treated wastewater is common in California, (e.g., Boyle Engineering Corporation 1981, see appendix in Pettygrove and Asano 1985). One of the oldest (over 60 years old) effluent-irrigated sites is in Bakersfield, in the southern part of the San Joaquin Valley. The crops grown in this area include cotton, corn, barley, alfalfa, sorghum, wheat and permanent pasture.

The effluent used in the area contains 24 mg N/l, which is considered adequate to satisfy the requirements of most crops. Nitrogen overdose in cotton cultivation caused excessive vegetation and poor setting of bolls. Similar observations were reported by Crites (1974) who used sewage effluent having the following chemical analysis (in mg/l): N (mainly as NH_4) 28; P 6; Cl 60; SO_4 90; HCO_3 220; Ca 15; Mg 18; Na 112; and K 12. The BOD of the effluent was 150 mg/l, the TDS 380 mg/l, the B level 0.5 mg/l and the SAR 4.4. The crop yields ranged from slightly below to well above the county average. This was the result of suitable management combined with effluent storage, control of tail water, circulation of water to prevent ponding, correct farming operations and application of appropriate amounts of effluent. As the mean annual rainfall in the Bakersfield area is only 150 mm, a large quantity of effluent was required, resulting in a high level of added N, which adversely affected cotton yields. The N loading for all the crops was much greater than the normally added amount, but cotton is the most sensitive of the crops to excess N (Table 4.4).

Pettygrove and Asano (1985) found that the addition of N through the effluent during the latter part of the growing season resulted in a decrease in cotton yield.

Table 4.4 Nitrogen loadings applied to the Bakersfield fields by means of sewage-effluent irrigation in comparison with standard fertilization; N uptake and crop yields (Crites 1974)

Crop	Actual N loading rate from effluent	Typical commercial fertilizer rate	N uptake	Yield
	----------(kg ha^{-1} yr^{-1})	----------	----------	(Mg/ha)
Alfalfa	416	—	112–165	18
Barley	156	88–101	83	3.4–5.6
Corn	282	224–275	165	40–67
Cotton	310	112	112	0.67–0.9
Pasture grass	416	>112	165–275	—

This was rectified by applying reclaimed wastewater only during the early part of the growing season and changing to canal water for the rest of the season.

Another interesting example from California is the Irvine Ranch Water District, south of Los Angeles, which is also discussed by Pettygrove and Asano (1985). In this area, sewage effluent is subjected to primary, secondary (activated sludge) and advanced treatment (dual media sand filtration and extended chlorination) and used to irrigate various fruit trees and row crops. Storage is a major problem, particularly during the winter of wet years. The treated effluent contains high levels of total dissolved solids (800–1000 mg/l) and boron (0.4–0.8 mg/l), which is typical for other reclaimed wastewater in southern California. Soil samples from citrus orchards and other fields are analyzed regularly at the beginning and end of the irrigation season, to control possible accumulation of toxic elements. Observations have detected a rise in the soil in total dissolved solids, boron and other materials during the irrigation season and a reduction following winter rainfall. The average level of boron during the last 10 years (up to 1985) was about 0.06 mg/l, with concentrations reaching 1.25 mg B/l. However, no toxic effects have been detected in citrus trees.

A field study carried out in Castroville, California (Burau et al. 1987) tested the feasibility of using tertiary-treated municipal wastewater obtained from a new plant with a capacity of more than 110 000 m^3/day, to irrigate artichokes and food crops to be eaten raw. The study area located near Monterey Bay is short in irrigation water and exposed to intrusion of seawater into coastal aquifers.

A 5-year experiment tested 3 types of water, 4 levels of N and 6 crops. The water types were: (1) filtered secondary effluent (filtered effluent); (2) physically and chemically treated secondary effluent fulfilling California requirements for irrigation of food crops eaten raw (therefore named "title-22 water"); and (3) local well water.

Burau et al. (1987) concluded that there was no significant difference in the yield or quality of food crops (artichoke, celery, broccoli, lettuce and cauliflower) irrigated with the two effluents tested, or between the effluents and the local well water. Furthermore, there was no public health risk associated with the irrigation of vegetable crops to be eaten raw with the two treated effluents, and no soil or ground water degradation was observed.

4.2.3 New Mexico

For 33 years, secondary effluent from domestic wastewater (treated by trickling filters) has been used for irrigation in Roswell, New Mexico (Koerner and Haws 1979). The treated effluent, transported to the field via concerte-lined ditches, is applied to the soil by furrow irrigation. The local mean annual precipitation (305 mm) is greatly exceeded by the potential evapotranspiration (2440 mm). About 800 mm of irrigation water is applied to the study area from March through July to corn and grain sorghum, and from March through September to alfalfa. Two sites were compared to study the effects of long-term irrigation with treated effluent vs. irrigation with fresh water. The crops grown on both sites were as follows: 1976, corn; 1972 to 1975, alfalfa; and 1966 to 1972, corn or grain sorghum. The fertilizers added were 78 kg N ha^{-1} yr^{-1} for corn and sorghum and 38 kg P ha^{-1} yr^{-1} for alfalfa. The yields obtained were 40 Mg ha^{-1} yr^{-1} of

Table 4.5 Quality of sewage effluent and well water used for irrigation crops at Roswell, New Mexico during 1976/1977 seasons (Koerner and Haws 1979)

	Effluent		Well water	
Constituent	Range	Mean	Range	Mean
		(mg/l)		
Dissolved oxygen	<0.1–6.1	2.0	7.3–7.7	7.5
Alkalinity as $CaCO_3$	316–336	323	168–196	183
Dissolved solids	2008–2504	2195	2374–3027	2700
Suspended solids	18–74	42	28	28
Total solids	2041–2522	2237	2127	2727
BOD	6–122	42	<1–2	1
COD	135–206	172	19–44	29
Dissolved organic carbon	14–58	28	8.4	8.4
Soluble organic nitrogen	<0.1–36	11	0.5–1.6	1.0
NH_4-N	17–26	23	<0.1–0.1	<0.1
NO_3-N	0.1–3.7	1.6	0.5–2.3	1.6
Total P	5.2–9.5	8.0	0.01	0.01
Soluble P	4.8–8.0	6.8	0.01	0.01
Soluble Ortho-P	3.8–7.8	65	0.01	0.01
SO_4-S	518–917	660	713	713
Chloride	545–820	681	710–933	844
Boron	0.13–0.21	0.179	0.24–0.46	0.337
Calcium	124–262	192	122–334	226
Iron	<0.2–1.8	0.64	<0.2–0.6	0.43
Magnesium	59–86	67	53–92	74
Potassium	13.7–16.8	14.7	2.1–6.2	4.8
Sodium	350–559	466	380–424	398
Zinc	0.01–0.09	0.05	<0.01–0.07	0.03
		($\mu g/m^3$)		
Lindane	49.2–1250	588	1.3–72	26
2,4-D	<1.2–64	29	0.6–18.4	7.6
2,4,5 TP Silvex	<0.2–70.4	28.1	0.2–5.8	2.0

silage corn (70% moisture), about 5 Mg ha^{-1} yr^{-1} of sorghum grain and 14 Mg ha^{-1} yr^{-1} of alfalfa as hay.

A summary of the constituents of the two water sources used for irrigation during the 1976/1977 season is given in Table 4.5.

Koerner and Haws (1979) investigated the effect of the same two irrigation waters on the quality of groundwater (at 2–4 m depth), soil and plant. Their main findings were as follows:

1. After 33 years of irrigation with treated effluent, indicator viruses and fecal bacteria did not penetrate the soil and thus did not reach the underlying groundwater.
2. No significant amounts of trace elements were found in the groundwater. Metals such as Zn, Cu and Cr as well as B remained in the soil and were not taken up by the crops more than from wellwater-irrigated crops.
3. The concentration of available N and P in the soil was 50% higher, and the potential loss of N through denitrification and volatilization was 3–30% greater in effluent-irrigated than in well-water irrigated soil. The concentration of NO_3-N in groundwater increased as a result of irrigation with sewage effluent.
4. Phosphorus penetrated to the top 30 cm in the effluent-irrigated soil. Calculations have predicted that the capacity of the Roswell soil for removing P will enable the soil to be irrigated with effluent for more than 100 years.
5. No harmful concentrations of the herbicides and pesticides endrin, lindane, methoxychlor, toxaphene, 2,4-D and 2,4,5 TD silvex were detected in the effluent or groundwater. The concentration of lindane in the groundwater of the effluent-irrigated site was greater than in the well-water irrigated site, but the levels were well below the current maximum allowable concentration according to the National Interim Primary Drinking Water Regulation.

4.2.4 Mexico

One of the largest areas irrigated with sewage effluent is located in the No. 3 Mezquital Irrigation District in Mexico, in the southwestern part of the Hidalgo state, north of Mexico City (Sanchez Duron 1985). The mean annual rainfall of the area is 494 mm, irrigation is essential to maintain productive agriculture.

Sewage effluent has been used for irrigation in the Mezquital area since 1886; in 1967, the effluent transportation system reached a capacity of 200 m^3/s. The water used for irrigation comprises sewage effluent mixed with rainfall, runoff water and clean water flowing from local reservoirs. During recent years (1979 through 1980) water used for irrigation amounted to approximately $1.01-1.42 \times 10^9$ m^3. The water flowing in the irrigation canals has a turbidity level of 104.3 Nephlometer Turbidity Units (NTU) and a clear to greenish-yellow color. It contains no sediments and is odorless and has a pH of 7.05 and EC of 1.47 and occasionally reaching 2.25 dS/m. The level of dissolved solids is about 1060 mg/l, of grease 23 mg/l, of organic C 6.9 mg/l and of alkyl-benzyl-sulfonate about 15.5 mg/l. The concentrations of the main cations are (in mmol (+)/l) Ca 6;

Mg 1.5; Na 3.6; and K 7, and those of anions are (in mmol(−)/l) HCO_3 3.47; Cl 5.5; SO_4 2.4; NO_3 4.16; and CO_3 1.88. The salinity hazard according to the USDA classification is between medium and high and the SAR level is marginally sodic, requiring special management practices before the water can be used for irrigation.

Other potential problems encountered in the Mezquital plant are the presence of relatively large amounts of heavy metals and pathogenic organisms. Yet the heavy metals precipitate in the soil and the concentrations of soluble Mn, Fe, Cu and Zn remain within the permissible levels. Furthermore, the majority of the crops grown in the area are tolerant or semitolerant to the boron content present in the irrigation water.

Although most of the houses in the area have safe drinking water, and the rest obtain it from public faucets, and despite the fact that consumption of fresh vegetables and certain fruits have been restricted, intestinal infections have not been eliminated. It is apparent that this problem is not necessarily the result of sewage effluent irrigation, but is caused by the general sanitary situation in the area. Demographic growth and bacterial and chemical pollution of the water and soil, and the overgrazing of pastureland are also responsible for the environmental contamination in the area. However, because of the economic advantages obtained from the use of sewage effluent for irrigation (i.e., a higher standard of living for people living under poverty, increased land value and improved farm income), the endemic diseases are considered a matter of secondary importance by the farmers.

Sanchez Duron (1985) notes that irrigation with sewage effluent has improved over the last 10 years, except for pasture irrigation. The presence of organic matter and plant nutrients in the effluent is believed to improve the fertility of the shallow soils of the area.

The generally good drainage conditions, and the calcareous and light-to-medium textured soils have contributed to the achievement of prosperous and stable agriculture in the region. The irrigation management include applying excess water to the crops (especially to alfalfa), to leach the soil and to prevent salt buildup.

The main crops grown in the area are pasture, alfalfa, corn, oats, barley, beans and vegetables such as Italian squash, chili pepper and tomatoes.

4.2.5 Werribee, Australia

Much information on the use of sewage effluent has been obtained from the Werribee farm, located 35 km southwest of Melbourne, Victoria, Australia (McPherson 1979). The mean annual rainfall in the area is 488 mm, distributed almost evenly throughout the year, with slightly higher amounts falling in winter. Evapotranspiration greatly exceeds rainfall, reaching about 1460 mm per year. The Werribee Farm was established in 1897 as a result of a cholera epidemic in Melbourne. Its total area (in 1979) was almost 11 000 hectares. The average flow of raw sewage entering the Werribee system is 0.44 $Mm^3\ day^{-1}$, with a BOD_5 of

570 mg/l and TSS of 620 mg/l. Three methods of purification are used: land filtration, overland flow (grass filtration), and lagooning. The first method is used during the late spring, summer and autumn (October–April, in Victoria) when evaporation is high and irrigation is normally applied. Overland flow is employed during the period of low evaporation (May–September) when irrigation is not practiced. Lagooning (aerobic and anaerobic) is used all the year round during periods of peak daily flows and wet weather. The treated effluent leaving the system flows to Port Phillip Bay, on the banks of which lies the city of Melbourne.

The land filtration section of the Werribee Farm (about 4280 ha) utilizes mainly raw sewage and provides water for irrigation of pastures. Each pasture bay is irrigated for 1 or 2 days, followed by 5–8 days of drying and 10–14 days of livestock grazing (a 18–21 day cycle). The farm supports 22 000 beef cattle and 30 000 sheep, the sale of which covers some of the sewage purification costs. The pasture basins are either flat plots or graded areas (180 × 10 m), and are protected by banks to prevent overflow of the applied sewage into the collecting drainage system. An average of 100 mm of sewage is applied to the upper end of the pasture basins and distribution is by gravity. About 50–60% of the applied water percolates and is collected in 1.2–1.8-m ditches, while the rest is lost through evaporation and seepage into deeper soil layers. Most of the nutrients and heavy metals are removed during land filtration, and the amounts eliminated are as follows (in % of the total): BOD_5 99; suspended solids 98; organic C 94; N and P 95; Zn, Cu and Pb 95; Cr 90; Cd and Hg 85; and Ni 75.

A considerable increase in the organic C, N and P was detected in soil exposed to long-term treatment with sewage effluent. Most of the N accumulation was detected down to 23 cm below the soil surface and the buildup of P reached a depth of about 46 cm. In accordance with local regulations, all cattle that had grazed on the farm were slaughtered on leaving the area, and the meat was carefully inspected for infection. Meat was frequently tainted with the tapeworm *Taenia saginata*, which causes beef measles; the larvae can be killed by cooking or freezing. The number of infected carcasses was approximately 0.02% of the total, which is about half the state average (McPherson 1979). The health of the farm workers is as good as that of the general community and there have been no epidemics. Offensive odors sometimes occur, but rarely extend beyond the farm boundaries.

Studies at the Werribee Farm include assessment of forestry, winter- and spring-sown cereal crops, oilseed, summer grain and forage crops, vegetable rotation and varied pasture mixtures for protein extraction.

4.2.6 Frankston, Australia

The Frankston Vegetable Research Station is located at the southeast edge of Melbourne, near Port Phillip Bay. The mean annual rainfall is about 760 mm, while the potential evapotranspiration is around 1400 mm. The main objective of the study carried out by Kaddous et al. (1986) was to investigate the effect of irrigation with reclaimed water on vegetable crops, namely, lettuce, carrot, cabbage, celery, spinach and tomato. The crops, growing in a loamy sand soil, were sprinkler irrigated. The following irrigation treatments were used: (1) Well water plus the amount of inorganic fertilizer normally given by the farmers; (2) reclaimed water plus the same quantity of fertilizer as in Treatment 1; (3) well water applied until the last side-dressing stage, followed by reclaimed water till harvest plus complementary inorganic fertilizers; (4) reclaimed water till the last side-dressing stage, followed by well water till harvest plus complementary inorganic fertilizers; and (5) reclaimed water plus the same amount of inorganic fertilizers as for Treatment 3. The chemical properties of the reclaimed water and well water are shown in Table 4.6.

High yields with high quality were obtained from both reclaimed water and well water plus standard fertilization. The difference in yields among all treatments (for each crop) were usually less than 15%. A combined application of reclaimed water and fertilizer resulted in the addition of a relatively large amount of nutrients to the soil. The yield obtained form this treatment was greater, but resulted in a delay in crop maturity (3–10 days) in comparison with well-water irrigation. Very high efficiencies were recorded for N and P uptakes from the reclaimed water. The use of reclaimed water saved about 60, 33 and 40% in N, P and K inorganic fertilizers, respectively, and around 35% of the fertilizer costs. The level of B in the reclaimed water (0.5 mg/l) supplied all B needs of the crop without any adverse effects due to excess B. As the level of the other trace elements tested (Pb, Cd, Cu, Cr, Fe, Ni and Zn) in the reclaimed water was low (below the permissible level for continuous irrigation), no accumulation occurred in crop tissues, and consequently no adverse effect was detected.

Table 4.6 Chemical composition of well water and reclaimed water used in the Frankston vegetable test (Kaddous et al. 1986)

	Reclaimed water		Well water	
Property	Range	Mean	Range	Mean
	[mg/l (Except otherwise noted)]			
pH	6.9–8.0	7.8	8.1–9.0	8.9
EC (dS/m)	0.73–0.95	0.88	0.97–1.3	1.18
Total N	11.1–14.1	12.0	0.1–0.6	0.45
P	3.8–5	4.5	0.01–0.08	0.06
K	12.5–15.2	14.0	2.0–4.6	4.0
B	—	0.5	—	0
CO_3^{2-}	—	0	—	12.0
HCO_3^-	9.8–12.00	104.0	80.0–116.0	110.0
SO_4^{2-}	52.0–65.0	57.0	7.5–15.5	13.0
Cl^-	126.0–301.0	165.0	192.0–547.0	470.0
Ca^{2+}	7.0–34.0	22.0	12.5–27.0	23.0
Mg^{2+}	6.0–28.0	13.0	18.0–28.0	25.0
K^+	6.0–20.0	12.0	15.0–28.0	19.0
Na^+	78.0–245.0	160.0	195.0–347.0	283.0

4.2.7 Alberta, Canada

Bole and Bell (1978) studied the effect of irrigation with sewage effluent on the yield and chemical composition of the following forage crops – alfalfa (*Medicago sativa* L.) reed canary grass (*Phalaris arundinacea*, L.), brome grass (*Bromus inermis* Leyss), Altai wild rye (*Elymus augustus*, Trin) and tall wheat grass (*Agropyron elongatum* Beauv.). The study area is located in Alberta, Canada, the climate of which is arid with 350 mm annual precipitation. Freezing temperatures restrict growth to May–September. The soil is a brown chernozemic loamy sand. The municipal water used contained 8.6–23.9 mg NH_4-N/l, 0.4–3.0 mg NO_3-N/l, 3–5.8 mg P/l, about 15–80 mg K/l and a total salt level of 720–1250 mg/l.

Yields were increased by the addition of N fertilizer, or by doubling the quantity of effluent above that necessary to meet the evapotranspiration demand (about 500 mm/yr). As water was not a limiting factor to growth, the N level controlled plant yield. Unlike N, P application through the effluent exceeded plant uptake. Alfalfa was the most suitable forage crop at optimum effluent application. The absolute forage yields obtained in this study were above the average for the crop and greatly depended on the quantity of effluent applied. However, the rise in water table prevented the application of large quantities of effluent unless drainage was controlled. The data obtained can be used for developing suitable irrigation-fertilization management well adjusted to the local conditions.

4.2.8 Israel

Irrigation with sewage effluent is widely practised throughout Israel, from the subhumid Mediterranean areas in the north to the extremely arid regions in the south. Winter (November–March) is cool and wet, but the rest of the year is dry, and evapotranspiration greatly exceeds rainfall. Consequently, intensive agriculture depends on irrigation. Laboratory, greenhouse and field studies have produced data on the response of soils and crops to irrigation with sewage effluent. The sewage effluent used for crop irrigation varies according to its source, nature of the previous water use and the type of treatment applied before reuse. Noy and Feinmesser (1977) have tabulated the typical composition of domestic wastewater, originating from Lake Kinneret supply water (Table 4.7). The difference between the third and fifth columns of Table 4.7 represent the minerals picked up during municipal use.

An extensive soil survey was carried out in Kibbutz Zor'a to study the long-term effects (up to 18 years) of irrigating Rhodes grass and cotton with sewage effluent. The survey was followed by a 5-year experiment.

Zor'a is located about 25 km west of Jerusalem. The soils are mainly grumusol (vertisol), characterized by a high montmorillonitic clay content and a relatively uniform profile. The effluent-irrigated sites contain $CaCO_3$ (approx 0.15 kg/kg soil) and the pH is about 8. The climate is Mediterranean with a mean annual (winter) rainfall of about 450 mm and a dry, hot summer.

Table 4.7 The resultant composition of wastewater from Kinneret supply water and average mineral pickup (Noy and Feinmesser 1977)

Composition	Unit	Supply water (Lake Kinneret)	Mineral pickup (Average)	Resultant wastewater composition
Ca	mmol(+)/l	2.43	1.41	3.84
Mg	mmol(+)/l	2.19	2.54	4.73
Na	mmol(+)/l	4.72	3.63	8.35
K	mmol(+)/l	0.13	0.48	0.61
Cl	mmol(−)/l	6.77	2.32	9.09
HCO_3	mmol(−)/l	2.70	3.55	6.25
SO_4	mmol(−)/l	1.61	1.06	2.67
EC	dS/m	1.05	0.56	1.61
SAR		3.1	2.0	5.1

Table 4.8 Range of some chemical properties of sewage effluent and fresh water used in the Zor'a experiments (1975–1981)

Constituent	Unit	Sewage effluent	Fresh water
pH		7.3–8.4	7.6–8.3
EC	dS/m	1.4–2.24	0.76–1.85
Na	mmol(+)/l	6.5–16.1	3.2–9.6
K	mmol(+)/l	0.3–0.7	0.1–0.2
Ca	mmol(+)/l	1.9–6.7	1.5–7.5
Mg	mmol(+)/l	2.6–5.4	2.8–4.4
Cl	mmol(−)/l	7.0–14.0	3.0–9.9
HCO_3	mmol(−)/l	6.6–12.8	2.6–5.3
P	mg/l	10.0–17.0	—
Organic-N	mg/l	1.0–23.0	—
NH_4-N	mg/l	38.6–68.0	—
NO_3-N	mg/l	0–4.0	0–4.5
Total-N	mg/l	45.0–85.0	0–4.5
B	mg/l	0.22–0.4	0.07–0.22
SAR		3.6–8.6	1.9–5.4

The quality of sewage effluent used for irrigation was periodically analyzed from 1975 to 1981. The values given in Table 4.8 represent the quality of the sewage effluents and fresh water used in the Zor'a study. The concentrations of Na^+, Cl^- and HCO_3^-, and the SAR level were considerably greater in the effluent. The effluent employed in the Zor'a experiments also contained high levels of P, NH_4^+ and organic N.

The salinity level of the top soil layer (about 0–20 cm) in different sites of the Zor'a-surveyed area was somewhat higher in the effluent-irrigated area (Table 4.9). Normally, however, winter rain and irrigation water resulted in salt leaching to deeper layers as indicated by Fig. 4.1. There was a clear increase in both EC and SAR in the water collected by means of a microlysimeter buried

Table 4.9 Highest EC_e measured in top layer of differently irrigated fields surveyed in Zor'a

Recent crop	Irrigation	EC_e (dS/m)
Lolium clover	Sewage effluent (3 yr)	1.97
Rhodes grass	Sewage effluent (7 yr)	1.67
Rhodes grass	Sewage effluent (11 yr)	1.67
Rhodes grass	Sewage effluent (18 yr)	1.45
Cotton (site 1)	Sewage effluent (7 yr)	0.93
Cotton (site 2)	Sewage effluent (7 yr)	0.81
Cotton	Sewage effluent (1 yr)	0.93
Cotton (site 1)	National carrier water	0.58
Cotton (site 2)	National carrier water	0.70
Wheat	—	0.41

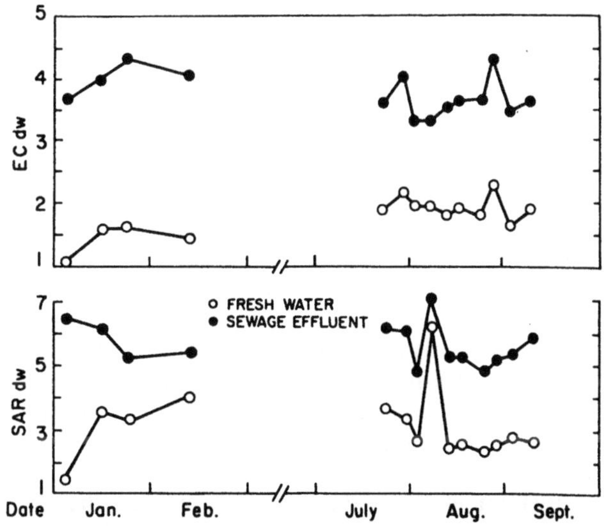

Fig. 4.1 Effect of irrigation with sewage effluent vs fresh water on the *EC* and *SAR* of drainage water collected by a lysimeter-like device using a suction pump (0.25 atm), situated 0.6 m below the soil surface of a clay soil planted with Rhodes grass in Zor'a (Vaisman, Feigin, Kipnis and Shalhevet, unpubl. data)

0.6 m below the surface of both an effluent- and well-water irrigated Rhodes grass field in Zor'a.

The ESP levels were considerably greater in the sewage-effluent treated soil after 3 years of irrigation (Fig. 4.2) and reached a steady state of around 6 after 7 years.

Table 4.10 summarizes a 5-year field experiment (Kipnis et al. 1981) carried out at Zor'a. The total quantity of N added by application of 1000 mm/yr effluent

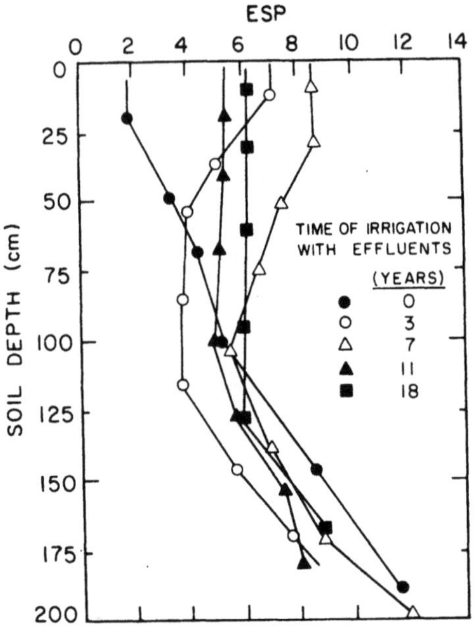

Fig. 4.2 Effect of irrigation with sewage effluent on ESP distribution in the soil profile of a few fields at Zor'a

was approx. 500 kg/ha and that of total P was more than 100 kg/ha^{-1} yr^{-1}. The yield of dry matter obtained from the effluent-irrigated soil was constant and greater than that measured for well-water irrigated plots on which the yield decreased greatly and steadily from 1975–1979. The quantity of N added through the effluent was nearly sufficient to satisfy plant needs, thus response of the grass to N fertilizer was small. On the other hand, a clear increase in Rhodes grass yield was obtained in response to N fertilizer when irrigation was with fresh water.

Bielorai et al. (1980) have shown that the cotton yield obtained for 3 consecutive years was greater in sewage effluent than in fresh-water irrigated plots. The great advantage of the effluent irrigation was evident in the no-fertilizer N treatment, and increased with time.

The suitability of drip irrigation technology for sewage-effluent application was studied in Zor'a during 3 consecutive years. Three effluent treatments (Bielorai et al. 1984), namely 350, 440 and 515 mm/season, were compared with similar quantities of fresh water to which N, P and K, in amounts similar to those present in the effluent, were added by fertigation. Two additional treatments were included in the experiment: 440 mm fresh water without additional fertilizer and with one-half of the full fertilizer treatment. The sewage effluent was filtered with standard filter used in drip irrigation systems, and no serious clogging problems were encountered. The yields obtained from effluent and the equivalent NPK fertilizer supplemented fresh water treatments were high and similar (Fig. 4.3).

Table 4.10 Effect of irrigation and N fertilization management on the yield of Rhodes grass obtained during a long-term field experiment located at Zor'a (Kipnis et al. 1981)

Treatment			Yield					
Water source	Irrigation frequency (times per week)	Fertilizer N ($kg\,ha^{-1}\,yr^{-1}$)	1975	1976	1977	1978	1979	Average yield
			(Mg/ha)					
Sewage effluent	1	0	18.1	15.5	17.2	16.0	15.0	16.4
		160	20.3	18.6	18.8	21.5	20.2	19.9
		320	20.6	19.3	22.7	26.3	21.9	22.2
		480	22.6	19.3	24.6	26.1	22.5	23.0
		Average	20.4	18.2	20.8	22.5	19.9	20.4
Sewage effluent	2	0	21.3	18.5	19.4	18.3	15.0	18.5
		160	21.4	18.9	22.2	21.7	18.4	20.5
		320	22.7	19.9	24.5	25.0	19.0	22.2
		480	21.7	20.7	24.6	25.4	18.4	22.2
		Average	21.8	19.5	22.7	22.6	17.7	20.9
Fresh water	1	0	—	14.1	4.9	2.0	1.9	5.7
		160	—	16.3	9.8	5.8	5.2	9.3
		320	—	17.7	13.2	9.0	11.4	12.8
		480	—	19.4	17.5	15.0	10.9	15.7
		Average	—	16.9	11.3	8.0	7.3	10.9

Soil and plant data (Feigin et al. 1981) indicated similar soil fertility levels in the effluent and the equivalent fresh water plus NPK fertilizer treatments. Excessive vegetative growth resulted in lodging and yield decrease during the first season, but appropriate agrótechnical improvement prevented such problems in the second and the third years (Bielorai et al. 1984) (see Sect. 5.4).

Two adjacent field experiments (designated A and B) were carried out at Tel-Adashim, located a few kilometers south of Nazareth, in the Yizre'el Valley (Dag et al. 1984). The local soil is grumusol (vertisol) rich in montmorillonitic clay (0.7 kg/kg soil), and contains 20–30 g $CaCO_3$/kg soil; the pH is about 7.7. The climatic conditions are generally similar to those in Zor'a (rainfall of approx. 500 mm/yr).

Two water qualities were tested: partially-treated municipal effluent (after sedimentation and oxidation pond treatment) and fresh water (Table 4.11). In both Tel-Adashim experiments (A and B), standard sprinkler irrigation scheduling (approx. 500 mm added in 3 or 4 irrigations during the growing season) were used. In experiment B the following irrigation treatments with

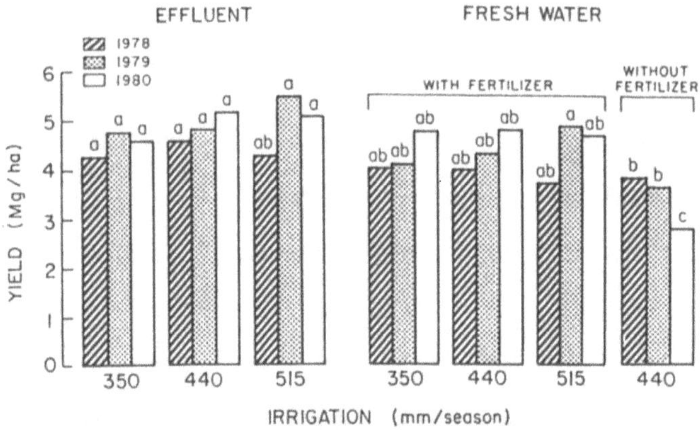

Fig. 4.3 Seed cotton yield in the effluent and freshwater-irrigated treatments at Zor'a during 3 years. Values of *columns* in the same year marked by *different letters* differ at the 5% level of significance according to Duncan's Multiple Range Test (Bielorai et al. 1984)

effluent were also included: (1) 20% more water than in the standard treatment; (2) about the same amount of effluent as in the standard treatment added in four or five irrigations, and (3) somewhat less effluent in two or three irrigations. In each of the two experiments the irrigation treatments were combined with four N-fertilizer treatments, added once as ammonium sulfate before crop planting. The fertilizer levels were as follows: (1) standard ($120\,kg\,N\,ha^{-1}\,yr^{-1}$); (2) half

Table 4.11 Average values and standard error of chemical properties of sewage effluent and fresh water used (1976–1982) in the Tel-Adashim experiments (Dag et al. 1984)

Constituent	Unit	Sewage effluent	Fresh water
	mg/l (except otherwise noted)		
pH		7.85 ± 0.25	7.52 ± 0.32
EC	dS/m	1.58 ± 0.19	1.14 ± 0.02
Na		200 ± 42	110 ± 30
K		28 ± 5	6 ± 3
Ca		104 ± 20	90 ± 26
Mg		47 ± 7	49 ± 14
Cl		302 ± 50	254 ± 64
Ortho-P		10 ± 3	0
Total-P		15.9 ± 4.3	0
NH_4-N		45 ± 8	0
NO_3-N		1	5 ± 2
Total-N		56 ± 9	5 ± 2
B		0.4 ± 0.02	0.05
COD		373 ± 94	—
SAR	$(mmol/l)^{1/2}$	4.3 ± 1.1	2.4 ± 0.7

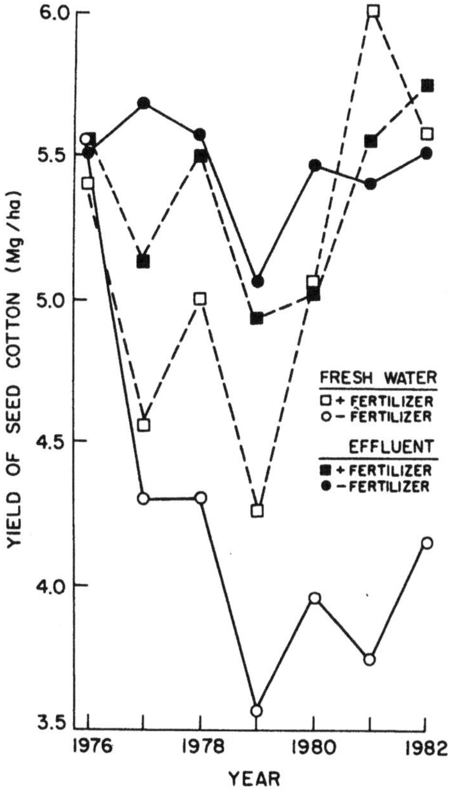

Fig. 4.4 Effect of irrigation with sewage effluent vs fresh water with and without fertilizer on the yield of seed cotton during 7 consecutive seasons in Tel Adashim (Dag et al. 1984)

standard; (3) one-and-a-half times the standard; and (4) no N fertilizer. Phosphorus fertilizer was added only once during the experimental period.

In experiment A cotton was grown for 7 consecutive years; in experiment B the first crop planted was sugar beet (1975/1976), followed by cotton (cv. SJ-2) grown for 5 consecutive seasons (1977–1981), after which wheat (1981/1982) and corn (1982) were grown. The yields of all crops irrigated with fresh water responded well to the addition of N-fertilizer. The high yields obtained in these experiments involved uptake of large quantities of available N from the soil, and fertilization was essential to keep soil fertility at the required level. There was only a small response to fertilizer when effluent containing 56 mg N/l was used (Fig. 4.4). Soil tests showed an improvement in soil fertility using sewage effluent (Table 4.12).

The effect of the method and frequency of irrigation on soil and crop (cotton) was studied in the eastern part of the Yizre'el Valley, using a food-industry-originated effluent having high salt levels (Cl concentration of 650 to 700 mg/l), a high Na content (SAR approx. 8) and containing large quantities of organic matter (comprising most of the 200–1500 mg/l suspended solids). The BOD of

Table 4.12 Effect of irrigation with sewage effluent vs fresh water, in combination with N fertilizer on the level of available N, P and K in Tel-Adashim soil (Feigin, Dag and Giskin, unpubl. data)

Soil Depth (cm)	Effluent			Well Water		
	NO_3-N	P	K	NO_3-N	P	K
	mg/kg Air-day soil					
0–20	14	12	6	9	10	5
20–40	14	12	5	9	4	7
40–60	14			9		
60–90	18			5		
90–120	15			4		
120–150	13			3		

the effluent was very high (approx. 2000 mg/l). Furrow irrigation was used commercially before the experiment in order to prevent leaf burn.

Despite winter rainfall both the salinity and sodicity levels in the soil greatly increased. Cotton yields were about 30% lower than the regional average for well water-irrigated cotton. Four irrigation treatments were tested as possible means for restoring the physicochemical condition of the soil and improving crop yields (Ravina, unpublished data). The five-hectare experimental plots were subjected to: 3-day-interval drip irrigation (Treatment 1); irrigation with moving spray machines (1600 l/h and 60-m boom length) at 7- or 14-day intervals (Treatments 2 and 3); and irrigation with a movable trickle-irrigation machine at 7-day intervals (Treatment 4). In the spring, before the start of the irrigation season, the EC_e values of the 0–0.6, 0.6–0.9 and the > 0.9 m soil layers were 1.5, 4.5 and 7.5 ds/m, respectively. The SAR values were approx. 7.5 $(mmol/l)^{1/2}$ in the upper soil layer and increased in the deeper soil layers [e.g., 14 $(mmol/l)^{1/2}$ in the 0.9–1.2 m layer].

The EC_e and SAR values increased following the commencement (in May) of irrigation. A steady-state was subsequently attained, with the EC_e values reaching 4, 6 and 7.5 dS/m for the 0–0.3, 0.3–0.6 and >0.6 m soil layers, respectively. Very high EC_e values were found beneath the cotton plants grown under drip irrigation (Treatment 1) and even higher in Treatment 4 (trickle irrigation machine) and the SAR of the 0–0.9 m layer increased to approx. 20 $(mmol/l)^{1/2}$.

The highest cotton yield was obtained from the 7-day-interval spray irrigation (Treatment 2), being only 10% smaller than the regional average, and no leaf burn was observed. A 5% lower yield was obtained from the similar Treatment 3, with a 14-day interval. A further 10% yield reduction was noted using the trickle irrigation machine (Treatment 4), while the lowest seed cotton yield was observed in the drip-irrigated field (Treatment 1). A gradual reduction in infiltration rate was observed in the soil subject to Treatment 1, and leaf yellowing appeared near to the peak of cotton plant growth, indicating poor soil aeration. One may attribute the yield reduction obtained from the 3-day-interval drip irrigation (Treatment 1) to the poor physicochemical conditions (high salinity and a high ESR), coupled with a high BOD level. Improved aeration obtained from the greater irrigation interval (Treatment 2) was probably the reason for the higher yield obtained with this treatment.

Trickle irrigation has been successfully used to irrigate cotton under arid conditions (annual rainfall approx. 200 mm). A good yield of seed cotton (6000 kg/ha) was obtained in a sandy loam soil following a seasonal effluent irrigation of 590 mm (Oron et al. 1982, 1987). (Wheat, alfalfa and corn were also successfully grown in the area.) As a result of proper screen filtration, no emitter clogging occurred. The total COD of the effluent was 160–220 mg/l, the BOD 20–40 mg/l and the ammonia level 20–75 mg/l. The total NH_3-N added, in the various irrigation treatments, was 180–210 kg/ha and the total quantity of orthophosphate P was 80–120 kg/ha; consequently, no additional fertilizer was necessary. The highest yield (about 6000 kg/ha) was obtained when the trickle irrigation was applied every 2 days using 2 l/h emitters, 1.92-m row spacing and 1-m spacing along the laterals. Twelve-day irrigation intervals, especially in the case of very wide lateral spacing (2.88 m), resulted in much lower yields.

Subsurface drip irrigation has been recently tested in the same vicinity (Oron et al. 1988). Good yields of alfalfa and cotton have been obtained during 7 consecutive years and no specific technical problem has been detected throughout the study.

Experiments on sprinkler irrigation of Rhodes grass and corn with treated sewage effluent were carried out in sandy dunes in the coastal plain of Israel. As shown by Kipnis et al. (1981), a large yield of Rhodes grass dry matter can be obtained from plants grown in sand dunes, as long as an appropriate irrigation-fertilization management is employed. Leaching of N below the root zone was a problem, and a careful irrigation-fertilization-cropping management was necessary to limit this process (Vaisman et al. 1981, 1982).

The yield of corn grown for fodder production in sand dunes varied greatly during 3 consecutive years, from a high yield of approx. 18 Mg dry matter/ha) (Vaisman et al. 1986) to a low yield of 9 Mg/ha (Vaisman, unpublished data). This means that much more information is still needed on different aspects of the appropriate cropping-irrigation-fertilization management and the protection of groundwater quality.

4.3 Humind Regions

Irrigation with sewage effluent has been extensively practiced in humid regions, mainly as a means of water treatment and disposal. Nevertheless, positive crop responses to effluent irrigation were obtained as a result of the additional water received by the crop during the relatively short dry periods which occur even under humid conditions. Response was also to the increase in the level of available nutrients resulting from effluent application.

4.3.1 Pennsylvania

An intensive long-term study on the "living-filter" approach was carried out in Pennsylvania State University (Parizek et al. 1967; Kardos 1967; Kardos et al. 1974, 1977). The yield of corn grain and silage responded positively to effluent irrigation. The addition of 50 mm treated sewage effluent/week resulted in higher yields than those obtained from plants irrigated with 25 mm effluent/week. Kardos et al. (1974) presented data (Table 4.13) showing that nutrient uptake (especially of N) was responsible for this result. The authors found that an effluent application of 50 mm/week throughout the year did not supply reed canary grass with sufficient nitrogen. Certain irrigation treatments did not supply K in amounts similar to those removed by crops; e.g., 50 mm effluent/week were able to replenish K removal by corn, but not that removed by alfalfa. The added quantities of effluent supplied more P than is taken up by crops in most years, and the amounts of Ca, Mg, B, Na and Cl were higher than those removed by the crops tested in the studies (Kardos et al. 1977). About 10% of the Cu and Zn added with the effluent was removed by silage corn and reed canary grass, and the latter crop also removed a large quantity of Cl. Some of the soil data obtained are discussed in Sect. 3.4.

4.3.2 Florida

Another example of effluent irrigation experiments aimed at both improved crop yields and renovation of groundwater is a study carried out in Florida (Overman 1975; Overman and Nguy 1975; Overman and Ku 1976). Tests have been carried out in Tallahassee since 1966 to evaluate the effectiveness of forage crops as a component of the effluent treatment process. Forage crops were chosen for this study since they are characterized by a high nutrient uptake. One of the crops

Table 4.13 Removal of sewage effluent constituents by means of corn silage irrigated with 25 and 50 mm of sewage effluent/week in the Pennsylvania "living filter" studies[a] (Kardos et al. 1977)

	1966				1969			
	25 mm		50 mm		25 mm		50 mm	
Element	kg/ha	%	kg/ha	%	kg/ha	%	kg/ha	%
N	134	115	156	78	285	286	218	120
P	30	65	42	51	33	178	65	123
K	104	66	145	49	159	264	107	97
Ca	30	13	34	8	31	33	18	7
Mg	20	13	24	9	24	35	20	15
Cl	40	9.5	49	6.2	53	22	37	10
B	0.05	2.0	0.07	1.0	0.09	1.0	0.05	0.5
Na	2.4	0.8	2.7	0.4	2.2	6.5	2.0	1.8

[a] Values higher than 100% indicate uptake values greater than thosse added through the effluent.

tested was pearl millet, [*Pennisetum typhoides* (Burm) Stapf and E.C. Hubbard; Overman 1975]. The experimental area was located near the wastewater treatment plant at Tallahassee. The crop was grown on a fine sand and the ground water depth was about 12 m. Sprinkler irrigation was applied at the rate of 13 mm/h and the quantities of effluent added were 50, 100, 150 and 200 mm/week. An irrigation rate of 50–100 mm/week appeared suitable for nutrient removal, the efficiency of which was reduced as the quantity of effluent was increased.

The response of corn (*Zea mays*), sorghum- Sudan grass (*Sorghum vulgare* pers. x *Sorghum sudanense* straf.) and kenaf (*Hybiscus cannabinus* L.) was studied in another experiment (Overman and Nguy 1975). The application rates of sewage effluent were 50, 100, 150 and 200 mm/week. The uptake efficiency of nutrients by the forage crops decreased with increasing effluent levels. The yield of corn was about 20 Mg/ha in the 200 cm/week treatment, compared to about 5 Mg/ha for the lowest irrigation application. The corresponding N uptake rates were about 250 and 100 kg/ha respectively. Similar trends in yield and N uptake were found for the other crops.

Rye grass (*Lolium multiflorum*) was also included in the Florida study (Overman and Ku 1976), with effluent application rates of 25, 50, 75 and 100 mm/week. The addition of 25 mm effluent/week was found to be suitable for growing both of the test crops in well drained soils. This Florida study showed that good yield can be obtained in sandy soils irrigated with sewage effluents, which provide plants with the essential nutrients.

4.3.3 Braunschweig, Germany

Valuable information on land treatment of sewage effluent under temperate climatic conditions had been obtained in Braunschweig, Germany. Ewert, Fassbender, Kayser and Tietjen, as well as other authors, quoted by AWWA (1981), presented and discussed relevant crop, soil, environmental and management aspects. A sewage farm, taking the Berlin farm as an example, was established at Braunschweig in 1896. The farm, designed for the disposal of 20 000 m^3 of sewage effluent per day, consisted of small beds, the total area of which was 300 ha, irrigated by means of surface flooding. In 1954, a new system was established to take care of a much greater volume of effluents, which resulted from the increasing local population. Beside the daily transportation of 15 000 m^3 to the sewage farm, 30 000 m^3 effluent per day were utilized by a few hundred farmers for crop irrigation. A net area of 3000 ha was irrigated 6 times a year with a total amount of 300 mm sewage effluent. Infiltration beds having a total area of 140 ha were also prepared to enable sewage effluent disposal during strong frost conditions. The cropland area was irrigated by means of sprinklers, specially designed for effluents (having a nozzle diameter of 10–15 mm). Crop yields obtained in the effluent irrigated area were higher than in the past, probably due to the additional water. A water deficit occurs every 2 years in the Braunschweig area, and once in 10 years it is exceptionally high (greater than 230 mm in the May–July period, and more than 260 mm between July–September). Soil fertility level had also been increased as a result of the effluent irrigation.

Ewert (AWWA 1981) mentioned surplus of water in the winter and excessive N in the spring as problems. Occurrence of offensive odors was also discussed and appropriate measures, such as improved aeration, were suggested to solve this problem.

Chapter 5 Irrigation and Fertilization Management

5.1 Site Selection – Soil, Hydrological and Climatic Characteristics

Proper site selection can reduce or prevent pollution problems (e.g., contamination of groundwater) and contribute greatly to a better use of effluent water and nutrients. The general criteria in selecting acceptable sites for land treatment of wastes are minimization of migration or transportation of pollutants from the place of waste deposit and maximizing pollutant biodegradation, retention and stabilization (Fuller and Warrick 1985). Addition of large quantities of low quality wastewater (primary or secondary effluents), and all the more so where sewage sludge is applied, should follow appropriate planning operations, site and crop selection included.

Site properties affect the planning, design and management operations necessary for proper utilization of sewage effluent for irrigation. Despite the large differences in the methodology and the principles of high-rate vs low-rate application of effluent, the main soil and hydrologic characteristics to be considered before establishing either type of system, are in principle the same. The objectives of the two approaches to effluent application differ, however, in the direction of water flow. Generally, in the high-rate application method, most of the water recharges the groundwater, whereas in low-rate application, most of the water evaporates.

The main site selection considerations are topography, soils, geology, groundwater and climate. The selection of sites for wastewater and sludge application was discussed by Hall et al. (1976), Fuller and Tucker (1977), Witty and Flach (1977), Crites (1985) and Fuller and Warrick (1985), among others.

5.1.1 Topography

The topographic parameters which are important in the selection and design of an irrigation system are slope and relief. A steep slope increases runoff and erosion and causes difficulties in cultivation. Relief, which is the difference in elevation over a land surface, affects pumping demands and water distribution uniformity. Susceptibility to flooding should also be considered. A major harmful effect of flooding is impaired soil aeration. However, flood-prone areas are generally deep, alluvial soils that, under a suitable irrigation method, careful management and flood protection, are suitable for sewage effluent application.

The topography of the surrounding area affects drainage and may also influence groundwater seepage into the site.

5.1.2 Soil Properties

Physical Properties. Texture, structure and soil depth are the main relevant physical parameters. The hydraulic behavior of soils is determined to a large extent by their texture. Fine-textured soils do not generally drain well; consequently, they retain large quantities of water for long periods, which enhances development of anaerobic conditions. Medium-textured soils are characterized by relative good drainage and are therefore suitable for irrigation with sewage effluents. Sandy soils need frequent irrigation, thus reducing water storage requirement.

Adequate and stable soil structure ensures proper drainage and aeration. Unstable soil structure may result in crust formation followed by a severe reduction in infiltration rate. Spray irrigation with large water drops, and alternate application of water having high and low salinity and SAR levels (e.g., cycles of sewage effluent irrigation followed by rain) may result in crusting. Certain medium-textured soils (e.g., loess and loessial soils) are especially prone to crust formation.

A soil depth of 0.6–0.9 m is considered as adequate for effluent irrigation, since it is deep enough to allow normal root development and the residence time of the wastewater in the "active" soil zone is sufficiently long. Information on soil profile properties is very important.

Hydraulic Properties. The design of an irrigation system and its operation should take into account the infiltration rate (IR) and the hydraulic conductivity (HC) or permeability of the soil. The hydraulic conductivity is the rate of water movement through the soil profile of an irrigated field by gravity (the hydraulic gradient being approximately equal to unity). The HC varies with the texture of the soil. The saturated HC of sandy soils is characterized by high values (10^{-2} cm/sec), and of clay soils, by low values (10^{-4}–10^{-5} cm/sec). Soils with a saturated HC of $<10^{-5}$ cm/sec (0.01 in/h) are considered impermeable. The HC may vary within the soil profile but the downward water flux is determined by the layer having the lowest HC. When the water application rate is smaller than the saturated HC of the soil profile, the soil will not become saturated, and its water content, as well as its HC, will depend on the application rate.

The infiltration rate – the rate at which water enters the ground – depends on soil texture and structure, initial water content and the rate of water application. At high application rates, or when ponded water enters the ground, the IR for a certain soil varies inversely with the initial water quantity, and decreases with time, approaching a steady-state minimum rate close to the soil saturated HC (Fig. 5.1). When the IR is lower than the application rate, ponding and runoff occur. Typical IR values for different soils are given in Table 5.1. It is recommended that the IR as a function of time be measured on test field plots

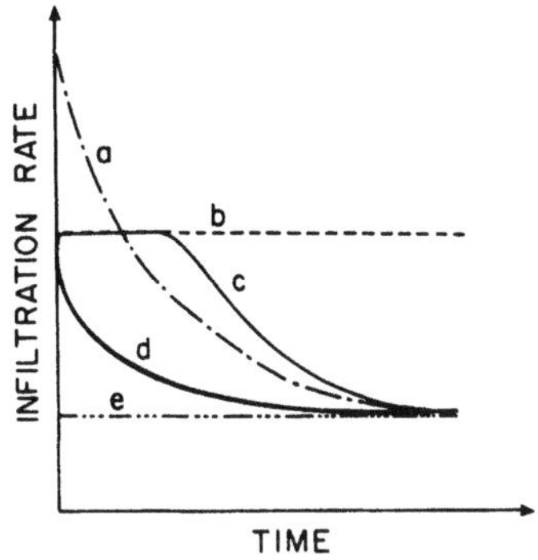

Fig. 5.1 Infiltration rate (*IR*) as a function of time under irrigation with a constant application rate or from ponded water. *a* Ponded water; *b* irrigation application rate; *c* dry soil; *d* wet soil; *e* saturated hydraulic conductivity

Table 5.1 Typical steady infiltration rate (IR) of various soil types

Soil type	IR (mm/h)
Sand	>20
Sandy loam	15–10
Silty loam	10–6.5
Clays	7.5–2

irrigated with effluent. Soils having an IR <10 mm/h are suitable for surface irrigation, while those with an IR of 10–50 mm/h or higher are suitable for sprinkler irrigation at either low or high application rate. Drip irrigation may be suitable in both cases.

Chemical Characteristics. The soil chemical characteristics influencing site selection for effluent irrigation include pH, electrical conductivity (EC_e), exchangeable sodium percentage (ESP) and cation exchange capacity (CEC). Other relevant factors to be taken into account are organic matter, boron and $CaCO_3$ content. At an ESP >10% with low salinity irrigation water, clay dispersion may enhance crusting, followed by a reduction in the infiltration rate and poor aeration.

The HC of the soil depends on the levels of Na and total salt of the percolation solution. A high HC can be maintained even at high ESP levels if the salt concentration of the percolating solution is above the critical level (Fig. 3.4). Clay dispersion occurs at low salt concentration. The negative effect of Na can be reduced or prevented by the presence of sufficient electrolytes released to the soil through dissolution of primary minerals or lime (Shainberg and Letey 1984).

5.1.3 Geologic Factors

The geology of the underlying strata affects groundwater contamination, especially where a shallow soil overlies a sandy subsoil. Fractured rock and geologic discontinuities may facilitate transport of percolating effluent.

5.1.4 Groundwater Depth

Groundwater depth is an important parameter in assessing the agronomic suitability of certain sites for effluent irrigation, as well as in drainage design and irrigation management. When a shallow groundwater is utilized for potable water supply, special precautions must be exercised to prevent pollutants from effluent irrigation from reaching the groundwater. From the adequacy of drainage point of view, groundwater depths greater than 2.5 m indicate the existence of good drainage conditions. In general, a groundwater depth of 1–1.5 m is adequate for most agricultural crops. Drainage may be necessary at shallower groundwater levels, especially where leaching for salinity control is involved. Alleviation of salinity occurs where saline water is mixed with sewage effluent having a low salt content (Day et al. 1981; Sabol et al. 1987).

Certain pollutants originating from sewage effluents are either not retained or are not inactivated by the soil, and the movement of others depend on soil properties. Dilution of such material in the groundwater can reduce the severity of the problem. Reliable information on the direction and rate of flow of aquifer groundwater is necessary for estimating the degree of dilution. Data on the location, depth and number of nearby wells are also required to estimate the extent of pollution problems. The upper layers of the water table will be more polluted than the deeper water layers. In a recent report by CAST (1985) it was suggested that NO_3 and bacterial contamination can be avoided by drilling farm wells deeper. In addition, the polluted zone tends to elongate in the direction of groundwater flow.

5.1.5 Climatic Factors

Rain, temperature, evapotranspiration and wind intensity and direction are important factors in crop selection and irrigation management. Wind parameters are also considered in the evaluation of air pollution from sprinkler spray. Climatic parameters necessary for planning irrigation management are specified in Sect. 5.2.

5.1.6 Practical Recommendations

Hall et al. (1976) suggested the following site criteria as ideal for both wastewater and sludge utilization:

1. Landscape – a closed or modified closed drainage system, preventing pollutant from wastes from reaching adjacent fields, and a slope smaller than 4% (unless a coarse-textured soil is involved) to avoid erosion. Open drainage systems, common under humid conditions, allows the transport of sediments and solutes from certain sites into water courses, and eventually to streams and rivers. In contrast, closed drainage systems, typical of dry climates, do not allow free flow of materials to the outside and, thus prevent pollution of areas outside their perimeters. It is possible to block open drainage systems (modified closed systems) by heaping small ridges across basin outlets.

   Fuller and Warrick (1985) suggest that wastewater application should be limited to lands of slopes smaller than 6%, while Crites (1985) recommends a maximum slope of 15% for cultivated agriculture and 15–20%, or greater, if local conditions allow, for noncultivated fields (e.g., pasture). Crites mentions that woodland areas having steep slopes (15–30%, and even more) were also successfully irrigated by sprinklers. Drip irrigation is especially suitable for such conditions.

   The presence of impermeable horizons in the soil (e.g., clay pans, shales, lime hardpans, silica cemented hardpans) restrict water movement to deeper soil or subsoil layers and become a problem where renovation of groundwater is the goal.
2. Parent material – a medium-textured soil material with a relatively high pH (6.5–8.2) and/or high level of free carbonates. Bedrock or unconsolidated strata (when present) should be free of coarse layers and should be at least 0.9–1.2 m below the soil surface.
3. Soils – a high infiltration capacity for the upper soil and a moderate subsoil permeability, no restrictive layer to a depth of at least 0.75 m, with good drainage conditions (lesser drained soils are acceptable when adequately tilled) and a moderate to high water holding capacity (15–20% by volume).

Less ideal sites may be used for wastewater and sludge application to support profitable crops where large land areas are available, provided that appropriate design and management measures are employed. Good yields of vegetable crops were obtained on sand dunes in the coastal plain of Israel when high-frequency irrigation was employed and an adequate supply of nutrients was added (Bar-Yosef 1977). Although fresh water was used in this study, sewage effluents may be applied (to suitable crops) provided that groundwater pollution is prevented.

Information on soils, geology, hydrology and topography as well as site visits and tests are conductive to successful planning of the system. Soil survey maps and equivalent data can be extremely valuable in the choice of sites suitable for effluent irrigation. In situ investigations by professionals (soil scientists, hydrologists and geologists) are necessary to obtain good results.

Tucker and Warrick (1985) suggested the following five-step procedure for site selection for wastewater and sludge disposal:

1. Reconnaissance for acceptable land and initial screening. Unacceptable sites are morains, fractured geological formation, high water tables, wet land, pH <5.5, shallow soils and similar situations.
2. Rating of soil characteristics according to physical and chemical properties.
3. Evaluation of landscape topography (slope, erosion susceptibility, land use, land forms).
4. Rating of hydrological and geographical factors: depth of bedrock or hardpan, groundwater quality, depth to residual soil, karst, topography, water sources, flooding frequency and wetland.
5. Final evaluation using information obtained in the previous steps.

5.2 Irrigation Scheduling

One of the important aspects of effluent water management is the availability of reliable information on crop irrigation water requirements. Municipal effluent is continuously supplied, regardless of irrigation needs, and it must be disposed of. If sufficient storage facilities are not available, irrigation must be adjusted, at least partially, to water supply. Furthermore, effluent may contain potentially hazardous constituents, which may cause soil and groundwater pollution, and may also directly affect plant growth. Consequently, it is critical that the quantity of water applied does not exceed the consumptive use and leaching requirements of the irrigated crop.

Sound water management requires information on appropriate irrigation timing relative to irrigation method, soil and root characteristics and crop development.

5.2.1 Crop Water Requirements

Crop water requirement may be defined as "the depth of water needed to meet the water loss through evapotranspiration (ET) of a healthy crop, grown under nonrestrictive conditions and achieving full production potential" (Doorenbos and Pruit 1977). It is the minimum amount of water which results in maximum yield in the crop-water production function, which relates yield to ET.

Estimation Based on Direct Measurements (Water Balance Approach). Field Plot Experiments. The best way of obtaining reliable water requirement values for a crop under a specific set of environmental conditions and cultural practices is by evaluating its water production function. Such functions may be obtained empirically by applying irrigation treatments with various known water quantities (I) to the crop in field plots, and measuring yield and changes in soil water content, before and after each irrigation (Shalhevet et al. 1981). Changes in soil water content are a result of evaporation from the soil surface and water used by the crop (evapotranspiration, E_c) and deep percolation (drainage, D).

The water balance equation may be expressed as

$$I = E_c + D \pm S, \tag{5.1}$$

where S in this case is the residual soil water content (+) or water consumed from storage in the soil (−). It is impossible in a field situation to directly measure D. However, under well-controlled experimental conditions, D can be minimized by normal wetting of the soil to less than full rooting depth. Such practice allows deeper roots to gradually deplete soil water stored deep in the root-zone and, thus provide room for slow drainage from the soil layers above. When deficit irrigation is practised, i.e., when less water is applied than that required to meet the water loss through ET, water loss through deep percolation is negligible in any case. Drainage (D) may also be estimated from the soil-water flow equation. This requires measurements of field unsaturated soil hydraulic conductivity and metric (tensiometric) gradients. Such estimates are very uncertain.

Line Source (Continuous Variable). The line source method (Hanks et al. 1976) may be used instead of field plots to obtain production functions. This method utilizes a single sprinkler lateral with sprinklers closely spaced along the line to provide uniform water distribution. The amount of water applied decreases as a function of increasing distance from the lateral. Yield and the quantity of water which falls on the crop canopy are measured at constant intervals from the lateral. This method is simpler and much less costly than large field plots.

Other Methods. Drip irrigation may be employed instead of sprinkler irrigation, using either small plots or single drip laterals placed along crop rows, with each lateral applying a different, measured volume of water. With gravity irrigation, small field plots are necessary.

Example of Direct Measurements. Once a complete crop-water production function is available in the range from deficit to excess irrigation, the water requirement can be estimated at the point on the production curve where maximum yield is obtained and further increase in water application does not result in further increase in yield (the yield plateau). Naturally, the excess quantities of water are not part of the ET curve. This water is lost through deep percolation.

Table 5.2 presents values of irrigation water requirements for some important crops in Israel, which were obtained by the use of production functions (Shalhevet et al. 1981). Similar tables have been published for other parts of the world by Doorenbos and Kassam (1979). Shalhevet et al. (1981) obtained production functions for cotton by graphically relating yield to the amount of water applied by irrigation, rather than to ET. The difference between the two functions, assuming uniform water application and negligible deep percolation, as was the case in the cotton experiments, is in the utilization of water stored in the soil

Table 5.2 Irrigation requirements, number of irrigations, irrigation interval and expected yields of field and orchard crops

Crop	Region	Irrigation requirements[a] (mm)	Seasonal irrigations (No.)	Irrigation interval days	Expected yield[b] (Mg/ha)
Wheat	Northern Negev and Lakhish	400–450	2		5.40
	Bet Shean Valley and Jordan Rift	500–550	2		5.00
Sorghum	Coastal Plain	250–280	2		9.90
	Northern Negev	380–420	3		7.80
Corn (grain)	Coastal Plain	330–350	4–6	10–14	8.50
	Northern Negev and Jezreel Valley	450–480	4–6	10–14	6.80
Cotton (lint)	Coastal Plain and Upper Galilee	330–360	3	21	1.76
	Northern Negev and Eastern Jezreel Valley	510–540	4–6	14–21	1.65
	Bet Shean Valley	780–810	6–7	14–18	1.70
Sugarbeets					
V[c]	All regions	70–120	0–1	21–30	10.80
VI	All regions	120–200	1–2	21–30	11.20
VII	All regions	180–250	2–3	21–30	11.70
Peanuts	Coastal Plain	530–560	5–6	18–21	5.10
	Northern Negev	530–560	7–8	12–14	4.60
Alfalfa (dry matter)	Coastal Plain and Upper Galilee	650–680	1–2	Between cuttings	20.00
Tomato (processing)	Coastal Plain	220–240	2–3	12–24	55.00
	Northern Negev and Lakhish	410–430	7–8	7	64.70

Citrus:					
Shamouti on sour orange	Coastal plain	600–660	8–10	20–25	47.50
Shamouti on sweet lime	Coastal Plain	500–570	8–10	20–25	63.00
Grapefruit on sour orange	Northern Coastal Plain	560–600	6–7	20–25	54.00
	Western Jezreel Valley	650–700	8–10	20–25	49.40
	Northern Negev and eastern Jezreel Valley	800–860	8–10	20–25	74.50
Avocado:					
Fuerte	Coastal Plain	700–750	10–12	14–21	13.70
Haas	Coastal Plain	750–800	10–12	14–21	22.50
Apples:					
Grand Alexander	Upper Galilee	760	12	10–14	118.00
Banana	Coastal Plain	1120	30	5–7	55.00

[a] Net water application for field crops, gross water application for tomato, alfalfa and orchard crops. Values for wheat include rainfall.
[b] 95% of the maximum yield.
[c] V, VI, VII are May, June and July harvests.

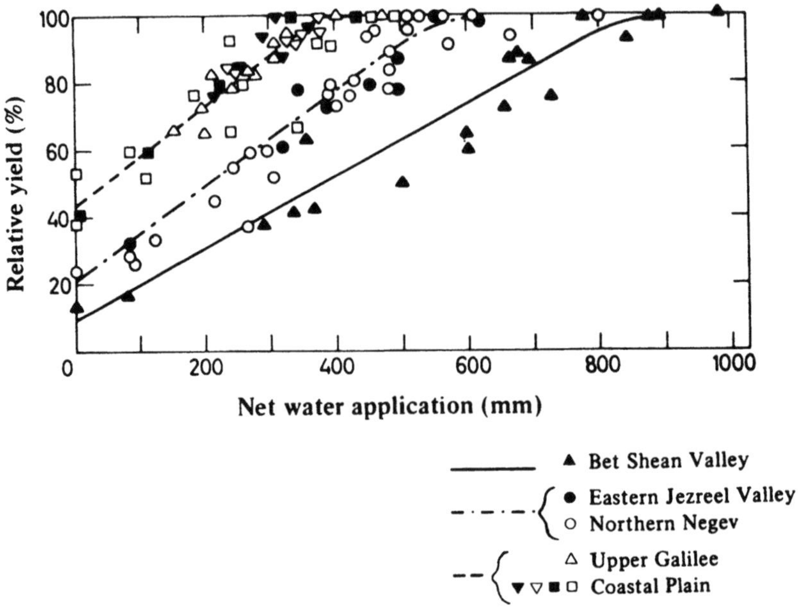

Fig. 5.2 Relation between relative cotton lint yield and net water application for five regions in Israel (Shalhevet et al. 1981)

before the irrigation season. A graphical representation of production functions based on water application is given in Fig. 5.2.

Use of Lysimeters. Another direct method is the use of drainage or weighing lysimeters placed in large fields with a single crop. Weighing lysimeters provide a closed system in which all the components of the water balance (E_c, D, I and S) can be measured. If properly designed and placed, they provide the most precise estimates of ET and are normally used to calibrate indirect methods of estimating ET. Lysimeters usually provide values for water requirement rather than production function, unless many such lysimeters are available.

Estimation Based on Climatological Factors. The dominant factor controlling the loss of water from crop and soil surfaces (evapotranspiration) is climate. The important climatic variables are radiation, temperature, humidity and wind speed. The influence of climate on ET is responsible for the three distinct production functions depicted in Fig. 5.2. The Israeli coastal plain is characterized by moderate summer temperatures (25–27 °C; mean values are given for the month of August) and high relative humidities (67–69%); the central region has higher temperatures (27–28 °C) and lower humidities (60%); the inland valleys are hot (31 °C) and dry (50% RH). In Israel, the rainfall varies between 500–600 mm along the coast to 300–400 mm in the inland valleys. Radiation does not vary among the three zones. These differences resulted in lower yield without irrigation (dryland) and in milder curve slopes (Fig. 5.2) in the inland valleys than along the coast.

The potential ET (E_p) in a given area may be computed from measurements of climatic variables. The potential ET commonly termed "reference crop ET" represents the upper limit or maximum ET that occurs under given climatic conditions within a field having a well-watered crop. Such a crop is defined either as an alfalfa crop 30–45 cm high with an aerodynamically rough surface (Jensen 1974) or an extensive surface of green grass crop 8–15 cm high, actively growing and completely shading the ground (Doorenbos and Pruit 1977).

There are numerous approaches for estimating E_p from climatological data. The approaches increase in complexity as greater accuracy is desired, by taking into account as many of the relevant climatic elements as desired. Each of the approaches was developed for specific climatic conditions and may not necessarily apply with the same degree of accuracy, if at all, under different climates.

To use the values calculated from climatic variables, they need to be converted to crop ET (E_c) by using a crop factor K_c. Thus $E_c = K_c E_p$.

The crop factor depends on the following characteristics of the crop and the climate:

1. Time of planting in relation to prevailing climate
2. Phenological stages of crop development, especially length of period to full ground cover
3. Length of growing season (number of days from planting to harvesting). Table 5.3 gives some data for various regions of the world regarding time of planting and length of growing season (Shalhevet and Kamburov 1973)
4. Leaf characteristics (stomatal density and opening, wax cover on leaves, etc.)

In any specific region, E_c for each crop must be calibrated against E_p by determining K_c for specific time intervals during the growing season.

In this section, the discussion will be limited to the simplest climatic approaches for estimating ET. The more complex ones, the so-called combination equations, especially the modified Penman equation (Penman 1963) give better estimates under most conditions and may be used to directly compute E_c, but require input variables which are not always readily available. Only a basic, simplified discussion will be given in order to acquaint the reader with the approaches. The interested reader is referred to Doorenbos and Kassam (1979), Jensen (1974, 1980), and Burman et al. (1980).

USWB Class-A Evaporation Pan. The simplest indirect approach is to measure evaporation from a body of water. Most widely used is the U.S. Weather Bureau Class-A evaporation pan. For the measurement to be applicable, the pan must be standardized and properly sited near the area of interest. Despite some obvious differences between pan evaporation and crop evapotranspiration conditions, such as reflectivity of solar radiation, heat storage in the pan during the day which

Table 5.3 Sowing and harvesting dates and consumptive water use of important crop plants

Crop	Country	Sowing date	Harvesting date	Consumptive use[a] (mm)
Wheat	Egypt	Oct–Nov	Jun	250–370
	Australia	May–Jun	Nov–Dec	350–500
	Canada	May	Aug	550
	Colombia	March–May	Aug–Dec	250
	India	Oct–Nov	Apr–May	
	Iraq	Oct–Nov	May	400–480
	Israel	Oct–Nov	May–Jul	370–420
	Pakistan	Oct–Nov	Apr–May	330–590
	Turkey	Oct	Jun	400–450
	USSR	Fall (winter)	Jul	100–200
		Apr (spring)	Jul	
Rice	Egypt	Apr–Jun	Oct	3700
	Australia	Sep–Oct	Apr–May	1600–1700
	Colombia	Jan–Mar	May–Jun	1500
	India	Jun–Jul	Oct–Nov	
	Iran	May–Jun	Aug–Sep	880–1240
	Pakistan	Jun–Jul	Oct–Nov	1280–1720
	Rep. of China	Jul	Nov	
	Turkey	Mar–May	Sep–Oct	
Corn	Egypt	Apr–May	Sep–Oct	
	Australia	Sep–Oct	Apr–May	500–635
	Colombia	Feb–Apr	Sep–Oct	470
	Iran	May	Aug	960–1090
	Israel (forage)	Apr–Jul	Jul–Oct	300–400
	Tunisia (forage)	Apr–May	Oct–Nov	450
	Turkey (silage)	Apr–May	Aug–Sep	700–900
Seed cotton	Egypt	Mar	Sep	860
	Australia	Oct	Apr–May	500–635
	Colombia	Jul–Aug	Dec–Feb (littoral)	500
		Jun–Mar	Jul–Sep (inland)	
	Iran	Apr	Sep–Oct	600–1170
	Israel	Apr	Sep–Nov	350–820
	Pakistan	May–Jun	Dec	820–1200
	USSR	Apr	Sep	460
	USA (California)	Apr	Sep–Nov	760–830
Sugarbeets	Canada	Apr	Oct	
	Iran	Mar–Apr	Oct–Nov	760–1190
	Turkey	Apr–May	Aug–Sep	
	USA (California)	Apr–May	Aug–Sep	860–940
Berseem	Egypt	Oct–Nov	May	625
	Pakistan	Oct	May	1030–1430
	Tunisia	Sep	Nov	850–1150
Sunflower	Australia	Oct–Nov	Mar–Apr	450–600
	Bulgaria	Mar	Oct	
	Iran	Jan–May	Jul–Dec	600
	Turkey	Apr–May	Aug–Sep	1260–2060

Table 5.3 *(Continued)*

Table 5.3 (*Continued*)

Crop	Country	Sowing date	Harvesting date	Consumptive use[a] (mm)
Sugarcane	Egypt			5000
	Colombia	Whole year	Whole year	2000
	Pakistan	Mar	Jan	
	Iran	Aug–Oct	14–19 months	3000
Alfalfa	Australia	May	Oct–Apr	1000–1800
	Canada	Spring		650
	Iran	Apr	Apr–Aug	880–2100
	Israel	Fall or spring	Summer	850–950
	Tunisia	Mar	Oct–Nov	1400–1600
Oranges	Israel		Nov–Apr	550–750
	Iran		Nov	960

[a] Figures in this column may include total water application rather than consumptive use per se.

induces night evaporation and wind effect, the Class-A pan provides reasonable estimates of ET. Calibration of pan evaporation (E_o) against crop ET (E_c) and determination of pan factor (K_o) are required for each specific area. An example of the distribution of E_c, for the cotton crop described in Fig. 5.2, is given in Fig. 5.3. The pan factor was found applicable under a variety of climatic conditions, as shown in Fig. 5.4 for cotton (see also Fig. 5.2) as well as for sorghum and grapefruit (Shalhevet and Bielorai 1978).

Pan Description: The standard pan is 121 cm in diameter, 25.5 cm deep, made of galvanized iron, and mounted on a wooden open frame platform with its bottom 15 cm from the ground level. The pan must be level, with the water level between 5 cm and 7.5 cm below the rim. Usually a chicken wire screen is used to cover the pan to prevent animals from drinking the water. The water must be kept clear. The water level is measured with a hook gauge placed in a stilling well.

Pan Site: The pan must be placed in an open area with no crop taller than 1 m, within 50 m from the pan. Around the pan there should ideally be 10 cm of short grass, but many pans are placed on bare soil. The pan should be placed in an agricultural area with irrigated crops.

Jensen–Haise Method. The Jensen–Haise method (Jensen and Haise 1963) requires input data of solar radiation and air temperature. The basic equation is:

$$E_p = C_T(T - T_X)R_S, \tag{5.2}$$

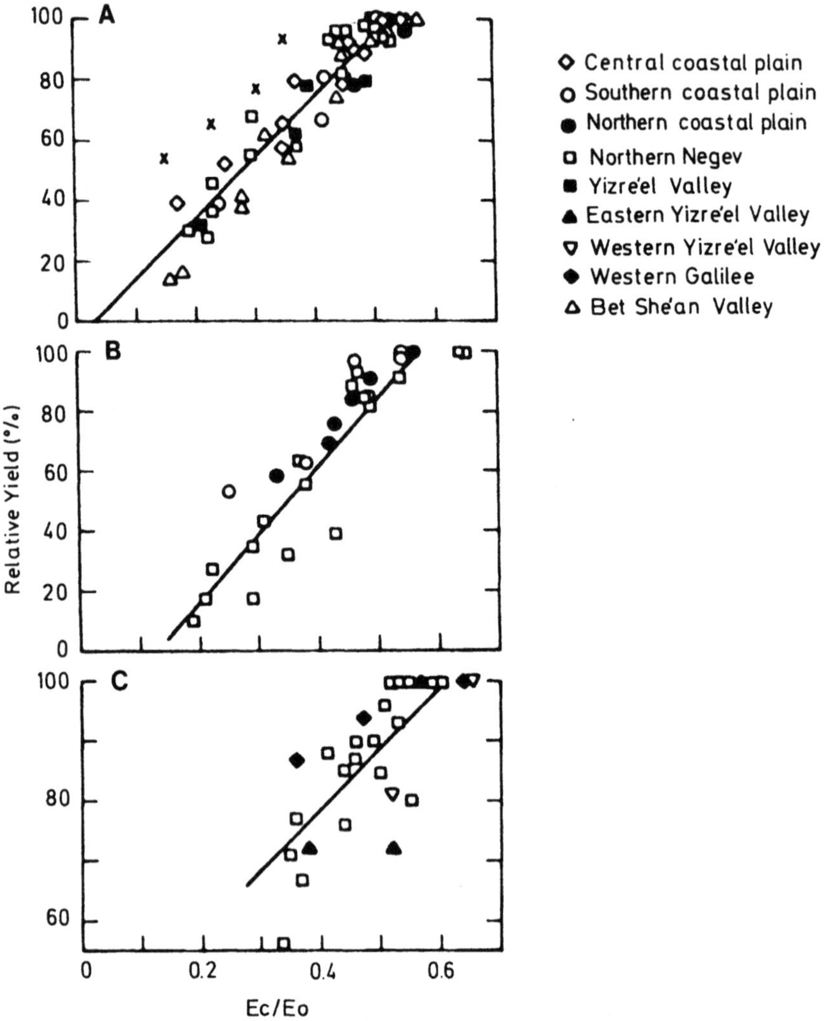

Fig. 5.3A–C Relationship of relative yield expressed in percent (*Y*), to the ratio of seasonal crop evapotranspiration to seasonal Class A pan evaporation (E_c/E_o) for (A) cotton, (B) sorghum, and (C) grapefruit for different regions in Israel (Shalhevet et al. 1981)

where E_p is given in mm/day, and R_S is solar radiation given in terms of mm of water obtained by using the latent heat of vaporization, L; ($R_S[\mathrm{cal\,cm^{-2}\,day^{-1}}] \times L\,[\mathrm{cm^3\,cal^{-1}}] = R_S[\mathrm{cm^3\,cm^{-2}\,day^{-1}}]$). T is the average temperature in °C for the period for which E_p is desired (shortest period 5 days). C_T is calculated by means of Eq. (5.3):

$$C_T = (C_1 + 7.3\,C_H)^{-1}, \tag{5.3}$$

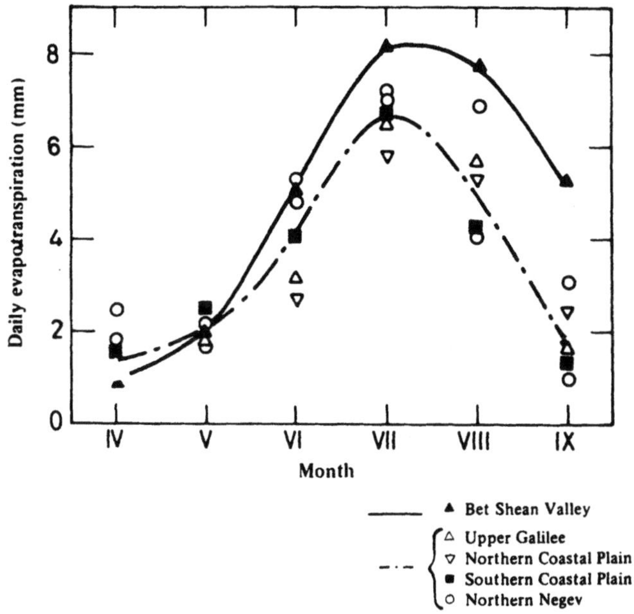

Fig. 5.4 Average daily evapotranspiration (E_c) by cotton in different regions in Israel (Shalhevet et al. 1981)

and C_H is obtained through Eq. (5.4):

$$C_H = 50(e_2 - e_1)^{-1}, \tag{5.4}$$

where 50, in mbar (mb), is a constant, e_1 and e_2 are the saturation vapor pressure of water, in mb, at the mean minimum and mean maximum air temperature of the warmest month of the year, respectively (long-term averages). C_1 is defined as follows:

$$C_1 = 38 - 2E(305)^{-1}, \tag{5.5}$$

where E is the elevation of the site above sea level, in m, and 305, in m, is a constant.

$$T_X = -2.5 - 0.14(e_2 - e_1) - E(550)^{-1}. \tag{5.6}$$

Thus, all unknowns in the above equations except T and R_s are site constants. R_S, if not available locally, can be estimated from Table 5.4 and from sunshine data, as described by Jensen (1974, 1980).

As with the pan evaporation method, the Jensen–Haise method must also be calibrated for E_c by determining the crop constant for each region. The yield response factors (K_c) for different crops and growth stages (vegetative period, flowering period, yield formation and ripening) and for the total growing period are given in Table 5.5 (Doorenbos and Kassam 1979).

Table 5.4 Mean solar radiation for cloudless skies expressed in cal cm^{-2} day^{-1} (calculated from Budyko 1963, cited by Jensen 1974)

Latitude	Month: Jan	Feb	Mar	Apr	May	Jun	Jul	Aug	Sep	Oct	Nov	Dec
					Latitude °N							
60	58	152	319	533	671	763	690	539	377	197	87	35
55	100	219	377	558	690	780	706	577	430	252	133	74
50	155	290	429	617	716	790	729	616	480	313	193	126
45	216	365	477	650	729	797	748	648	527	371	260	190
40	284	432	529	677	742	800	755	674	567	426	323	248
35	345	496	568	700	742	800	761	697	603	474	380	313
30	403	549	600	713	742	793	755	703	637	519	437	371
25	455	595	629	720	742	780	745	703	660	561	486	423
20	500	634	652	720	726	760	729	697	680	597	537	474
15	545	673	671	713	706	733	706	684	697	623	580	519
10	584	701	681	707	684	700	681	665	707	648	617	565
5	623	722	690	700	652	663	645	645	710	665	650	606
0	652	740	694	680	623	627	616	623	707	684	680	619
					Latitude °S							
5	648	758	690	663	590	587	577	590	693	690	727	677
10	710	772	681	640	571	543	526	558	680	690	727	710
15	729	779	665	610	516	497	497	519	657	687	747	739
20	748	779	645	573	474	447	445	481	630	677	753	761
25	761	779	626	533	419	400	406	439	600	665	767	777
30	771	772	600	497	384	353	358	390	567	648	767	793
35	774	754	568	453	335	300	310	342	530	629	767	806
40	774	729	529	407	281	243	261	290	477	603	760	813
45	774	704	490	357	229	183	203	235	447	571	747	813
50	761	669	445	307	174	127	148	177	400	535	727	806
55	748	630	397	250	123	77	97	123	343	497	707	794
60	729	588	348	187	77	33	52	74	283	455	700	787

Table 5.5 Crop factor (k_c) for different crops and growth stages (Doorenbos and Kassam 1979)

Crop	Vegetative period			Flowering period	Yield formation	Ripening	Total growing period
	Early	Late	Total				
Alfalfa			0.7–1.1				0.7–1.1
Banana							1.2–1.35
Bean			0.2	1.1	0.75	0.2	1.15
Cabbage	0.2				0.45	0.6	0.95
Citrus							0.8–1.1
Cotton			0.2	0.5		0.25	0.85
Grape							0.85
Groundnut			0.2	0.8	0.6	0.2	0.7
Maize			0.4	1.5	0.5	0.2	1.25
Onion			0.45		0.8	0.3	1.1
Pea	0.2			0.9	0.7	0.2	1.15
Pepper							1.1
Potato	0.45	0.8			0.7	0.2	1.1
Safflower		0.3		0.55	0.6		0.8
Sorghum			0.2	0.55	0.45	0.2	0.9
Soybean			0.2	0.8	1.0		0.85
Sugar beet							
Beet							0.6–1.0
Sugar							0.7–1.1
Sugarcane			0.75		0.5	0.1	1.2
Sunflower	0.25	0.5		1.0	0.8		0.95
Tobacco	0.2	1.0				0.5	0.9
Tomato			0.4	1.1	0.8	0.4	1.05
Water melon	0.45	0.7		0.8	0.8	0.3	1.1
Wheat							
Winter			0.2	0.6	0.5		1.0
Spring			0.2	0.65	0.55		1.15

5.2.2 Determination of the Irrigation Interval

In the computation of the crop water requirement, it is assumed that the crop is well watered, i.e., that soil water is easily available. Soil water may be easily available over a wide range of irrigation intervals, depending on soil storage capacity, unsaturated hydraulic conductivity and rooting depth. When irrigation is applied by gravity and portable sprinkler methods, it is normally desirable to increase the irrigation interval to the maximum possible value, but that which is still compatible with the concept of "easily available soil water". With solid-set sprinkler systems and drip irrigation, water is normally aplied frequently.

A reasonable rule of thumb is to irrigate crops when no more than 50% of the available water (WA) in the root zone of crop has been depleted. Some crops such as wheat, sorghum and cotton may withstand greater depletion, up to 70–75%. Others like potato may require a smaller depletion of 40–45%. It should be born

Table 5.6 Water availability (WA) in some soil texture groups

Soil texture	WA in (cm/m) Range	Mean
Coarse sand	2–6	4
Sand	4–9	6
Loamy sand	6–12	9
Sandy loam	11–15	13
Fine sandy loam	14–18	16
Loam and silt loam	17–23	20
Clay loam	14–21	18
Clay and silty clay	13–18	15

Table 5.7 Depth of rooting of some important crops when root growth is unrestricted

Shallow roots 60 cm	Medium depth 100 cm	Deep 120–150 cm	Very deep 180 cm
Lettuce	Peas	Cotton	Alfalfa
Onion	Beans	Melon	Grain corn
Celery	Carrot	Sugarbeet	Tomato
Potato	Beet	Wheat	Safflower
Cabbage	Sweet corn	Peanuts	Deciduous trees
Cauliflower	Cucumber	Citrus	Grape vine
	Pepper		Olives

in mind that crops suffer from water deficit much before soil water content reaches the permanent wilting point throughout rooting depth.

Table 5.6 provides information on the amount of water available between field capacity and wilting percentage in soils of various textures, and Table 5.7 gives typical rooting depths of some crops. These tables provide only general guidelines. In some fine-textured (clay) soils the nature of the hydraulic conductivity-water content relationship is such that at a higher water content the availability of water is lower (water is held in the soil by stronger forces) than in medium-textured (loam) soils. This would require irrigation at shorter intervals than would be the case on the basis of volumetric water availability alone.

Some soils may have compact subsoil horizons, restricting root development to shallower depths than those given in Table 5.7. In such cases less water will be available than for deep-rooted crops.

5.3 Irrigation-Fertilization Interrelations

5.3.1 General Considerations

Irrigation with treated sewage effluent involves addition of both water and nutrients to the soil-plant system. Optimization of water application in accordance with economic and environmental considerations also affects crop nutrition, since the amounts of the different substances added are also influenced. The distribution of nutrients contributed by sewage effluents during the irrigation season also differs from that of nutrients supplied by fresh water amended by fertilizer (fertigation) since with effluent, nutrient quantities cannot be directly controlled. Furthermore, effluents containing considerable quantities of N may cause nutrient buildup in the soil during sensitive growth periods, such as those close to harvest time.

Appropriate irrigation-fertilization management enhances efficient use of nutrients by crops resulting in high yields, reduced expenses on fertilizer, decreased water pollution hazard and mitigation or prevention of physiological disorders due to excessive concentration of nutrients in the root zone (Hanks et al. 1983). The differences in the chemical properties of sewage effluent and fresh water should be reflected in the adoption of specific management operations. The differences are usually small in the case of advanced-treated effluents, and gradually increase as the level of treatment is reduced. The use of raw sewage necessitates major changes in management.

Standard fertilization recommendations can be used for sewage effluent irrigation if they are adapted in the light of data from relevant experiments and local farmers' experiences. This is practised successfully by Israeli farmers growing industrial, fodder and grain crops.

5.3.2 Nitrogen Management

The crucial role of N in the maintenance of soil fertility, in determining crop yield quality and in pollution of surface water and groundwater has led to N management becoming the subject of many studies and discussions. Numerous publications (e.g., Pratt et al. 1978; Keeney 1982a; Hauck 1984a, b; CAST 1985) deal with irrigated soils in general, while others discuss specific aspects of effluent-irrigated fields (Feigin et al. 1978; Smith and Peterson 1982; Linden et al. 1983; Broadbent and Reisenauer 1985).

Appropriate application of N to crops should be based on optimal economic returns, a minimum damage to the environment and an understanding of the water, crop, soil, climate and other components of the relevant system, supplemented by data from soil and plant analyses. The available N derived from sewage effluent replaces equivalent or similar amounts (depending on the case; see below) of fertilizer.

Different soil and plant analytical methods have been suggested for determining the actual quantities of fertilizer necessary under specific field

Table 5.8 Plant N concentration associated with deficiency, sufficiency and excess in several important crops (Olson and Kurtz 1982)

Crop	Form of N measured	Plant part	N content (g/kg dry matter) at designated nutritional status or crop			
			Deficient	Low	Sufficient	Possibly excessive
Alfalfa	Total N	Top 15 cm at early bloom	< 40	40–45	45–50	> 50
Bermuda grass	Total N	Whole tops 4–5 weeks after clipping	< 15	15–25	25–30	> 30
Brome grass	Total N	First cutting for hay at early flower stage	10–15	15–20	20–25	> 25
Corn	Total N	Ear leaf at silk	< 22.5	22.6–27.5	27.6–35	> 35
Cotton	Total N	Upper mature leaves	< 25	25–30	30–45	> 45
Grain sorghum	Total N	Third leaf below head at bloom stage	< 25	25–30	30–40	> 40
Potato	NO_3-N	Petiole of 4th leaf from growing tip	< 8	8–10	10–12	> 12
Millet	Total N	Whole tops 4–5 weeks after clipping	< 19	19–22	25–35	> 35
Sugar beet	NO_3-N	Petioles of recently matured leaves	< 1	1–2	2–3	> 3
Wheat (winter)	Total N	Total above-ground plant at head emergence from the boot	< 12.5	12.5–17.5	17.5–30	> 30
(Spring)			<15	15–20	20–30	>30

conditions (e.g., Walsh and Beaton 1973; Page et al. 1982; Hauck 1984b; Haynes 1986). Often fertilizer recommendations are prepared for specific conditions [e.g., Reisenauer (1978) for crops grown in California].

Plant tissue analysis, yielding the N concentration in specific plant parts, is used for fertilizer decision-making. This method helps to determine the nutritional status of the plant and, if necessary, indicates the time, quantity and mode of fertilizer application. Tissue analysis is widespread in perennial crops, where there is time for effective correction of nutrient deficiency, but is commonly used for evaluating the nutritional status of other agricultural crops (Table 5.8). The plant tissue for analysis and the time of sampling are usually very carefully selected, because there are large differences in the level of nutrient elements in different plant parts and the concentrations of these elements change during the growing season. As shown in Tables 5.8 and 5.9, leaves or petioles are used in most cases for this purpose.

Climatic factors, especially temperature and moisture, affect the nutrient level in plants as a result of its effect on dry matter production (Olson and Kurtz 1982). A typical change in the level of NO_3-N, PO_4-P and K in the petioles of cotton plants during the growing season is shown in Table 5.9. If the level of NO_3^--N is below the "low" level at first bloom, and the finding is confirmed by another test a week later, top dressing would be recommended (Bassett and McKenzie 1978).

Soil testing is a valuable tool for assessing N requirements under various conditions and is successfully used in effluent-irrigated fields. However, as the level of available N in soil can easily and quickly change, soil N data should be professionally evaluated. A major advantage of soil tests is that they provide information on fertilizer requirements prior to planting. Various chemical and biological soil testing methods have been suggested for predicting the potentially available N in the soil (Keeney 1982b; Goh and Haynes 1986). Recently, models using soil and meteorological data have also been proposed. Residual mineral N (mainly NO_3^--N) in the soil is similar in availability to standard fertilizer N. Under conditions of negligible N losses through leaching or denitrification, the level of NO_3^--N in soil near to crop planting time is considered as available for the crop.

Table 5.9 The level of NO_3-N, PO_4-P and K in cotton petioles at different sampling time (Bassett and MacKenzie 1978)

	NO_3-N		PO_4-P		K	
Sampling	Low	High	Low	High	Low	High
	(g/kg dry matter)					
First bloom	12	18	1.5	2	40	55
Peak bloom	3	7	1.2	1.5	30	40
First open ball	1.5	3.5	1	1.2	20	30
Late season	Should fall below 2		0.8	1	10	20

Table 5.10 Some biological N availability indices (Keeney 1982b)

Temperature (°C)	Time (Days)	N form	Related to
			Aerobic
35	14	Min N[a]	Sugarcane N uptake, field
30	14	Min N	Rye grass N uptake, greenhouse
25	21	Min N	Rye grass, barley, N uptake and response, greenhouse, field
26	40	Min N	Oat, N uptake, greenhouse
28	126	CO_2 release	Rye grass, maize, N uptake, greenhouse
35	14–210	Min N	Development of N mineralization potential, N_o
30	21	NO_3-N	Sugar beet yield, N uptake, petiole NO_3^--N in conjunction with profile NO_3^--N, field
			Anaerobic
30	14	NH_4-N[b]	Aerobic N mineralization
30	14	Min N	Rye grass, N uptake, greenhouse
40	7	NH_4-N	Rye grass, N uptake, greenhouse
35	14	NH_4-N	Development of N mineralization potential, N_o

[a] Min N = (NH_4^+ + NO_3^-) mineralized.
[b] NH_4-N = NH_4^+ mineralized.

Table 5.11 A range of chemical N availability indices (Goh and Haynes 1986)

Extractant	Temperature (°C)	Time (h)	Form of measured N
	Mild extractants		
Water	100	1	Total N
0.001 M $CaCl_2$	100	64	Total or NH_4^+-N
0.001 M $CaCl_2$	121	1	NH_4^+-N
0.01 M $NaHCO_3$	Room	0.25	Total N or UV absorbance
1 M KCl	100	1	NH_4^+-N and NO_3^-
2 M KCl	80	20	NH_4^+-N and NO_3^-
0.01 M $CaCl_2$	121	16	Soluble carbohydrate
0.1 N $Ba(OH)_2$	Room	0.5	Soluble carbohydrate
	Intermediate extractants		
Alkaline $KMnO_4$	100	0.25	NH_4^+-N
Neutral 0.5 N $Na_2P_2O_7$	100	6	NH_4^+-N
1 M NaOH	100	0.5	NH_4^+-N
1 M NaOH	Room	4.2	NH_4^+-N
1 M HCl	Room	26	NH_4^+-N
1 M H_2SO_4	Room	1	NH_4^+-N
	Strong extractants		
6 N H_2SO_4	Room	28	NH_4^+-N
4.5 M NaOH	NaOH distillation		NH_4^+-N
$K_2Cr_2O_7$	Walkley-Black Oxidation		NH_4^+-N

A second and/or complementary approach is the estimation of the "potentially mineralizable" soil N (Keeney 1982b; Goh and Haynes 1986). Such tests predict the quantity of N that becomes available during the crop growing season. These methods are divided into two main groups: biological and chemical. Some information on typical aerobic and anaerobic incubation tests is shown in Table 5.10, while general characteristics of chemical tests are given in Table 5.11. Detailed information on the procedure of recommended biological and chemical methods is given by Keeney (1982b). The selected method should be suited to the prevailing soil-plant conditions. Adequate information on plant N needs can also be obtained from field studies, as well as from the experience gained by farmers.

The N balance principle has been used for predicting the N fertilizer requirements of crops (Stanford 1973, 1981; Meisinger 1984). Meisinger (1984) discussed mass balance principles, and presented two different situations, differing in their physical boundaries. The "whole crop N balance" model, which includes within its boundaries both the above-ground portion of the plant and the root zone component of the soil, is illustrated in Fig. 5.5 and described by Eq. (5.7):

$$N_f + N_{misc} - N_{ch} - N_e - N_l - N_g = N_{son} + N_{sin}, \tag{5.7}$$

in which N_f is fertilizer nitrogen input, N_{misc} comprises various nitrogen sources (e.g., rainfall, irrigation or NH_3 adsorption), N_{ch} is N removed in harvested crop, N_e is nitrogen lost by erosion, N_l is nitrogen lost by leaching, and N_g is gaseous nitrogen losses (NH_3 volatilization and denitrification). The N_{Son} and N_{sin} are the changes in the soil organic and inorganic N content, respectively.

Another approach is the "above-ground crop N balance" (Meisinger 1984) in which the boundary of the systems is drawn at the soil surface (A'B' in Fig. 5.5). The fertilizer N input (N_f) is calculated using Eq. (5.8):

$$N_f = \frac{N_{ch} + N_{cr} - e_m N_{min} - e_s N_{sin}}{e_f}, \tag{5.8}$$

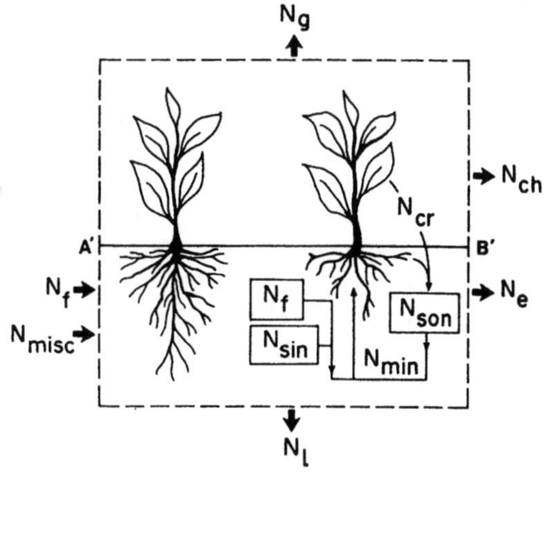

Fig. 5.5 Schematic representation of the N balance in a soil-crop system (Meisinger 1984)

in which N_{min} is estimated soil mineralization, N_{sin} is estimated soil mineral content, N_{ch} and N_{cr} are the nitrogen contents in the harvested and above-ground crop residues, respectively. The e_f, e_m and e_s are the efficiency terms of the respective fractions of N_f, N_{min} and N_{Sin} in the above-ground crop residues. By contrast to the "whole crop N balance approach" [Eq. (5.7)], the effects of soil transformations are only indirectly represented in the efficiency terms.

The above-ground N balance approach was used to evaluate N in soil-plant systems under different cropping systems (Parr 1973; Standford 1973, 1981; Carter et al. 1976). Feigin and Shakib (1972) employed this approach to determine the N fertilizer needs of processing tomatoes. The quantity of fertilizer to be added was based on a 100 Mg/ha fresh tomato yield having a total N content of 160 kg/ha, and on crop residue N content of 40 kg/ha. The quantity of available N (1 N KCl-extractable NO_3 and NH_4-N) in the root zone to a depth of 1.2 m was 60 kg/ha and the potentially mineralizable soil N, estimated (using previous data) on 2% of the total organic N, was 50 kg N/ha. The calculated N deficit of 90 kg N/ha (N required−N available) was to be added through fertilization. Since the utilization efficiency of fertilizer N added to tomato crops under similar conditions in previous experiments was about 70%, then 90/0.7 = approx. 130 kg/ha fertilizer N was added in Feigin and Shakib's study. This approach was also successfully employed in cotton fields, a major user of treated-sewage effluent in Israel. As both tomato and cotton are deep-rooted crops grown in well-controlled irrigated fields during a typical dry Mediterranean summer, losses through leaching and denitrification were relatively low (as indicated by a relatively high N uptake efficiency of 0.7). Reliable information on crop N requirement is essential for efficient fertilization. Such information can be obtained from relevant field studies and from appropriate analyses made in commercial fields.

Combined application of fertilizer and irrigation water (fertigation), especially via a drip system, can improve N uptake efficiency (Bresler 1977; Bar-Yosef 1977; Feigin et al. 1984). A major advantage of this method is the well-controlled supply of nutrients to the developing plants. Reliable information on the optimal nutrient level in the soil during the growing season can greatly improve nutrient uptake efficiency and crop yield. Such information is usually obtained from field trials. It is also possible to divide the growing season into a few major periods during which the required quantities of water and nutrients are supplied.

Usually, N uptake efficiency of effluent N is similar to that of standard fertilizers (such as ammonium sulphate), especially when effluent is used for extended periods. However, lower efficiency may occur when leaching is required to prevent salinity buildup or when denitrification is increased, as a result of reduced aeration caused by effluents with high COD or BOD levels. Losses through NH_3 volatilization may occur in alkaline soils. For instance, N uptake efficiency of sewage-effluent NH_4-N added to a clay soil in a pot experiment (Feigin et al. 1981) was about 15% lower than that of solid fertilizer [$(NH_4)_2SO_4$] carefully incorporated into the soil. Some of the effluent N has probably been lost through NH_3 volatilization, while the NH_4 added was apparently nitrified to NO_3 after its application, and then exposed to denitrification.

Nutrient in the effluent is added as long as irrigation water is applied and thus cannot be stopped during the crop ripening period. This disadvantage can sometimes be rectified by suitable management practices. For example, a moderate water stress can restore the sugar content of sugar beet, and prevent lodging and boll shedding in cotton (see Sects. 3.3 and 4.2). Reduction of cotton stand may facilitate development of thicker stems which also contribute to the prevention of lodging. The addition of fertilizer N to effluent-irrigated soil may sometimes be necessary, but excessive application must be prevented.

The distribution of available N in the soil prior to planting can affect fertilizer needs. Certain crops are efficient N scavengers and the level of available N left in the top soil, from which the subsequent crop takes up its essential N, is low. A similar situation of N deficiency can occur in fields exposed to heavy rainfalls. Additional N fertilization to replenish the available N in the top soil layer is sometimes necessary under such conditions, even though the total quantity of available N within the whole root zone is sufficient. It should also be noted that often a crop is planted into a wet soil, resulting in a delay of the first effluent irrigation and therefore in a delay in the application of N present in the effluent.

5.3.3 Phosphorus, Potassium, Calcium, Magnesium and Sulfate Management

Most data obtained from low-rate effluent application show an increase in soil P (Table 3.11), but do not indicate specific problems resulting from excessive P buildup. When the quantities of P added to the soil with the effluent satisfy crop needs, addition of fertilizer is unnecessary. Data on the critical level of available P for different crops is given in Table 5.12. Such information is essential for making fertilizer recommendation. Suitable data on tissue analysis is also available for many crops (e.g., Reisenauer 1978).

Assessment of the P status in the soil is usually done by conventional soil or plant testing (Olsen and Sommers 1982). These methods are also suitable for effluent irrigated soils. Water soluble, acid extractable (dilute $HCl+NH_4F$, dilute $HCl+H_2SO_4$; see Table 5.12), and 0.5 N $NaHCO_3$ extractable P are common. The latter method is especially satisfactory in calcareous, alkaline and neutral soils.

Soil testing may provide information on P requirement. The amount required in fertilizer P can be estimated by deducting the P added with the effluent from the total requirement.

Phosphorus applied in the effluent will concentrate in the top layer and will usually not be available until it is incorporated into the soil or is otherwise transported into the main root zone. Under drip irrigation, the roots of crops are concentrated within the continuously maintained wet zone where the concentration of P is high and is more readily available for the growing crop than under sprinkler irrigation.

Table 5.12 Critical soil P levels, (mg/kg soil) at which a response (in crop yield) to P fertilization is not likely[a]

Crop	Soil test	Region	P level[b]
Corn	$NaHCO_3$	California	>12
Cotton	$NaHCO_3$	California	>8
Cotton	$NaHCO_3$	Israel	>12
Pasture and forage	$NaHCO_3$	California	>10
Potato	$NaHCO_3$	California	>25
Potato	$NaHCO_3$	Israel	>24
Sorghum	$NaHCO_3$	California	>7
Sugar beet	$NaHCO_3$	California	>12
Tomatoes for processing	$NaHCO_3$	Israel	>15[c]
Wheat	$NaHCO_3$	Israel	>10
Wheat	$NaHCO_3$	India	11
Wheat	$NaHCO_3$	USA, calcareous soils	10
Wheat	$NaHCO_3$	Bolivia	7
Wheat	0.025 N HCl +0.03 N NH_4F	Mexico	25
Corn	0.025 N HCl +0.03 N NH_4F	Mexico	25
Corn and sorghum	0.025 N HCl +0.03 N NH_4F	USA	30
Millet	0.05 N HCl +0.025 N H_2SO_4	North Carolina:	
		Silty clay loam	10
		Fine sandy loam	17

[a] The California data are from Reisenauer (1978); the Israel data are from Halevy (1979); Halvey et al. (1975, 1987) and Feigin and Sagiv (1975); the other data are from Kamprath and Watson (1980).
[b] The level of no response is indicated by the (>) sign.
[c] In early planting (relatively cooler temperature) a higher value (approx. 30 mg P/kg air dry soil) was obtained.

Although chemically greatly different from P, K movement within the soil profile is also slow. Potassium availability to crops depends on its presence within the active root zone of the plant. As in the case of P, the quantity of K added in effluent can be deducted from that planned to be applied as fertilizer. Soil analysis is also a very useful tool to keep the level of available K within the recommended range.

Calcium and Mg are often present in large quantities in effluent. In addition to their positive effect on water quality related to soil physiochemical conditions, they are important nutrients. The level of Ca and Mg in calcareous soils is usually satisfactory and Ca and Mg fertilization is not required.

Sulfate is also important for plant growth, since S is one of the essential nutrients. The presence of SO_4 in effluent can be therefore advantageous in S deficient soils.

5.4 Trace Elements

Different aspects of trace element behavior in sludge- and sewage effluent-treated soil-plant systems were discussed in Sect. 3.6. Some practical considerations relevant to sewage effluent are presented below.

Due to their potential phytotoxic and food-chain effects in regard to the use of wastewater and sludge, certain metals (Cd, Pb, Hg, As, Se, Zn, Ni, Mo) are of special interest. Information on trace element content in sewage effluents must be obtained, and available recommendations concerning the use of trace elements in regular farming practices should be consulted. Data on the permissible levels of trace elements in water used for short term and long term irrigation are present in Table 2.2 and Sect. 3.6.

Application rates of sewage sludge depend on its trace element content and on the N requirement of crops. At low trace element levels, N becomes the limiting factor, and vice versa.

The maximum concentrations (mg/kg) of heavy metals in sludges considered acceptable on agricultural land by the Commission of the European Community are as follows: Cd 20–40; Cu 1000–1750; Hg 16–25; Ni 300–400; Pb 750–1200; and Zn 2500–4000; and the limit values (in kg ha^{-1} yr^{-1}) of heavy metals added annually to agricultural land are: Cd 0.15, Cu 12, Ni 3, Pb 15, Zn 30, and Hg 0.1 (Sauerbeck 1987).

Chaney and Giordano (1977; see also Logan and Chaney 1983) suggested the following permissible levels (given in mg/kg dry weight) of trace elements in sludge, that would be appropriate for land application, for Zn 2500; Pb 1000; Cu 1000; Ni 200; Hg 10; and Cd 25; the permissible Cd/Zn ratio is 0.01.

The quantity of Cd in sludge permitted by the US EPA (1979) should not exceed 0.5 kg ha^{-1} yr^{-1}.

Soil pH and its cation exchange capacity (CEC) affect trace elements toxicity, and must be considered when the permissible loads of trace elements are determined. The USEPA (EPA 1979) stated that the pH of soil/sludge mixture at the application time should be at least 6.5 where food-chain crops are grown. For lower trace element cumulative crops, less stringent regulations are used. Liming, usually given to increase soil pH in acid soils, also reduces the probability of trace element accumulation.

The impact of soil CEC on the permitted loading of trace elements is shown in Table 3.14 (see Sect. 3.6). Logan and Chaney (1983) mention the following problems as related to the use of CEC in trace element management: (1) it seems that Cd uptake by crops grown in calcareous soils is not sensitive to the CEC; (2) the effect of the CEC on the uptake of Cd may be different than its effect on other elements; (3) the effect of CEC on trace elements uptake seems to be most valid for soils having a mixture of organic matter, clay minerals and Al, Fe and Mn oxides, but not necessarily for soils predominated by any of these components.

5.5 Crop Selection

Various factors are involved in selecting crops for irrigation with sewage effluent. These include: economics, soil and crop characteristics, effluent properties, climate, equipment available, farmer's skill, labor and tradition. The most typical crop selection criteria are: legal and regulatory requirement, potential hazards to the crops (salinity, boron, chloride, etc.) and management requirements (George et al. 1985). The decisions to be made include variety selection, scheduling activities, preparation of seedbed, weed and pest control, fertilization, timing and application of irrigation, labor management and marketing of the crop.

There are reports of a variety of crops which have been successfully irrigated with sewage effluent (see Chap. 4). Many of the species were not tested extensively, while others were grown on a large scale. The geographic distribution of these crops depends to a large extent on soil and climatic conditions. Certain crops grown with sewage effluent were mainly used as scavengers in disposal sites, with the main aim to improve the quality of the renovated water. The yield obtained had an economic value, which covered part of the expenses involved in the renovation operations. In some cases the role played by the crops was of little importance, while in other cases it was major. In areas where the main purpose of irrigation with sewage effluent was to produce profitable yields, the effluent water was used in place of unavailable fresh water, and was often the only source of irrigation water.

Many publications discuss the uptake efficiency of elements by crops. Kardos et al. (1977) mentioned the successful use of corn for silage and grain, and alfalfa and reed canary grass for absorbing large quantities of nutrients while obtaining high yields of dry matter. The response of various forest trees to effluent irrigation was evaluated by Smith and Evans (1977); they noted that sewage effluent can contribute to renewing the wood fiber resources, while at the same time solving waste recycling problems. The employment of appropriate tree species and management practices are preconditions for success.

The principal crops of the cool subhumid and humid regions are fall-seeded grains, grasses and legumes; corn, for silage and grain, is also an important crop (Webber and Beauchamp 1977). Various species, especially of cotton and grain, were found suitable for the arid regions of the USA (Fuller and Tucker 1977). Wolcott and Cook (1976) mentioned reed canary grass and tall fescue (*Festuca elatior* L. var *arundinacea*) as most suitable plants for the formation of dense sod, under the climatic conditions of the north-central region of the USA, for high-rate effluent application. Timothy (*Phleum pratense* L.) was found to be a good species adapted to moist conditions, that could be grown together with ladino clover (*Trifolium repens* L.) or birdfood trefoil (*Lotus corniculatus*). Numerous crops (corn grain, corn silage, wheat grain, soybeans and grass, and forest trees) were evaluated as N scavengers. George et al. (1985) have recently reviewed the suitability of many crops for effluent irrigation, and discussed the effect of climatic conditions as well as other factors. Forage crops and turf grasses suitable for cool conditions include bluegrass, bromegrass, fescues, ryegrass, orchard grass, reed Canary grass, wheat grass, timothy, clover and trefoils. The warm season perennials mentioned in the list were Bermuda grass, St. Augustine grass, Dallis grass and Rhodes grass. The annual crops listed were winter cereals, and summer annuals: corn, Sudan grass and sorghum. Cotton is successfully grown within the required temperature limits. Woody perennials used for landscaping are also irrigated with sewage effluent. Rice also grows well under effluent irrigation.

Various grasses have been successfully grown in Australia (McPherson 1979) and vegetables were also tested where advanced-treated effluents were available (Kaddous et al. 1986; Burau et al. 1987). Many crops exhibited satisfactory growth in other parts of the world. Experiments have produced successful growth of cotton, sugar beet, corn for fodder and grain, wheat and Rhodes grass in Israel, provided suitable irrigation–fertilization–cropping management was followed (Feigin et al. 1978; Bielorai et al. 1984). As mentioned above, the main purpose of irrigation with sewage effluent in Israel is the production of profitable crops while utilizing the water resource efficiently.

5.6 Drainage

5.6.1 Introduction

To avoid accumulation of toxic constituents, be they nutrient elements, trace elements or salts, in the root zone, a net downward flux of water averaged over time must be maintained beyond the root zone (van Schilfgaarde 1984). Under most conditions, a satisfactory net downward flux is maintained naturally whenever the amount of irrigation water applied exceeds the amount used consumptively. In the presence of a shallow water table, a net downward flux may not be naturally available, in which case a flux must be provided through an artificial, subsurface drainage system.

A drainage system may also be necessary if the water leaving the root zone contains pollutants which must be prevented from reaching even deep groundwater. Such a drainage system may be functional only if a slowly permeable subsoil strata exists and a perched water table is built up. The prevention of pollutants from reaching the groundwater is necessary only if the groundwater is reused for municipal water supply or irrigation. When the aquifer water quality is initially poor, the groundwater may serve as a convenient sink for pollutants.

It is beyond the scope of this section to provide a detailed discussion of drainage of irrigated lands. For further information the reader is referred to Luthin (1965) and van Schilfgaarde (1974). The following provides a short description of drainage needs and design considerations when leaching of salts and other constituents of effluent water is necessary, and the upward movement of hazardous chemical constituents from a shallow water table is a problem. Artificial drainage may be required whenever the phreatic surface or the saturated zone above it (capillary fringe) is within the root zone for extended periods. When the groundwater is saline, which is often the case in arid regions, and when the irrigation water contains hazardous chemicals, the need for drainage may be even more acute.

5.6.2 Criteria for Drainage Needs and Design

A basic requirement for any artificial drainage system is the availability of an outlet for the disposal of the drainage water. Without an appropriate outlet, artificial drainage cannot be provided. An outlet must be at a lower elevation than the depth of the drainage system, so that drainage water may reach it by gravity flow. To reach an outlet, a pumping station must sometimes be provided. Drainage water, if of reasonable quality, may be reused for irrigation. However, eventually the steady, unavoidable deterioration in drainage water quality will necessitate an ultimate disposal site. Such a site may be a river, lake, marsh, sea or evaporation ponds. Canals or ditches normally convey the water to the ultimate disposal site. The disposal of drainage water from effluent irrigation is especially critical, as the water may contain high concentrations of salts, nitrogen and organics.

Three criteria influence the need for drainage and the design of a drainage system:

1. The tolerable depth to the water table and the tolerable duration of high water table
2. The drainage coefficient, i.e. the required rate of removal of excess water from the root zone
3. The hydraulic properties of the soil

Tolerable Water Table Depth. Two factors control the critical depth of the water table, i.e., the crop root distribution characteristics and aeration requirements, and the height and the rate of the upward salt movement from the water table. In the absence of salinity, a water table depth of 1.0–1.2 m is satisfactory for most crops (Wesseling 1974). For shallow-rooted crops like potatoes (see Table 5.7) the permitted depth may be less.

When the groundwater is saline, as is the case in most irrigated areas, especially when the irrigation water contains high concentrations of salts or other pollutants, the critical water table depth needs to be deeper. This depth will be controlled by soil water management and environmental characteristics and the salinity of the groundwater.

Talsma (1963) studied the upward movement of salts from the water table under a variety of conditions in Australia. From his results, and those of others, he concluded that the critical depth to the water table for the control of soil salinity is 1.0–2.0 m. The criterion for determining the critical depth is that depth which maintains the evaporation rate at less than 0.1 cm/day. At lower values, a large decrease in water table depth is necessary to change the flow rate appreciably. At shallower depths the evaporation rate increases rapidly as the water table rises. At an evaporation rate of 0.1 cm/day, salinization can be controlled. The critical depth depends on a combination of soil properties, but, in general, medium-textured soils are more sensitive than either sandy or clay soils.

The approach described above is applicable when irrigation intervals are relatively large (weeks), or for nonirrigated crops. Modern irrigation practices using sprinkler or drip methods favor short irrigation intervals. Under such management practices, a net downward flux may be more easily maintained. In fact, the question of critical water table depth for salinity control becomes secondary when a net downward flux is maintained. The determining factor then becomes the rooting and aeration requirements of the crops.

The above considerations pertain to a permanent position of the water table. For short periods following an irrigation, the water table may be allowed to rise to 0.3 m or less with no ill effects (Bouwer 1974), provided the rise is due to excess irrigation water of reasonable quality, not seepage nor a decreased depth of a saline groundwater table.

Drainage Coefficient. The drainage coefficient refers to the required rate of removal of excess water to maintain adequate drainage. This criterion is used in the design of drainage systems. Under steady state conditions, as is the case in many humid regions, the drainage coefficient will allow for sufficient drainage to maintain the water table at a desired depth. Under irrigation conditions (transient state), an adequate drainage coefficient will permit rapid lowering of the water table following irrigation.

Van Someren (Bouwer 1974) suggested a drainage coefficient of 3–7 mm/day for the conditions of the Netherlands, depending on the crop and the severity of the drainage problem. The generally accepted coefficient is 7 mm/day. Talsma (1963) suggested a coefficient of 0.5–2.5 mm/day for salinity control of irrigated fields.

The leaching requirement discussed previously (Sect. 3.1.2) can be used to estimate the required drainage coefficient as shown by Shalhevet (1973); e.g., a crop with a tolerable drainage water salinity C_d of 15 dS/m (a moderately sensitive crop) is irrigated with water of $EC_i = 2$ dS/m. If the seasonal consumptive use requirement of the crop is $D_e = 50$ cm, the amount of drainage water D_d will be:

$$D_d = \left(\frac{EC_i}{C_d - EC_i}\right) D_e = \left(\frac{T}{15-2}\right) 50 = 7.7\,\text{cm}. \tag{5.9}$$

If the water is applied in six irrigations of 9.6 cm each, then 1.3 cm of water will have to be drained away within a few days (5 or 6) following an irrigation, i.e., an average seasonal drainage coefficient of 0.25 cm/day must be maintained.

A certain salt accumulation may sometimes be allowed before leaching is necessary. If leaching is provided, for example, every third irrigation, the drainage coefficient will be increased to 0.75 cm/day, requiring a drainage system able to carry more water within 5 or 6 days following that irrigation. The choice of leaching procedure will therefore depend also on the cost of the drainage system.

The above considerations are based on the assumption that there is no rainfall during the irrigation season (Mediterranean-type climate). During the period between irrigation seasons, the required water table depth and drainage coefficient will depend on rainfall conditions, subsurface seepage and whether a crop is grown or not. When municipal effluent water is used and water storage facilities are not available, the water must be used for irrigation or otherwise disposed of. As pointed out above, the most efficient method of disposal is through irrigation of a crop, also during the rainy season. The amount of water which can be applied during the rainy season should not exceed the water requirements of the crop grown. More water can be applied than required for disposal purposes only when there is no need for artificial drainage (i.e., in the case of deep water tables and permeable soils).

Soil Hydraulic Properties. The two most important soil properties for drainage design are the hydraulic conductivity (K) and drainable porosity (p) of the soil. Given a specified drainage coefficient, the hydraulic conductivity will, to a large extent, determine drain spacings. The hydraulic conductivity of the soil is essentially the rate of saturated water flow in the soil under unit gradient. The actual rate of water movement towards drains will generally be considerably lower. The saturated hydraulic conductivity is generally higher for a sandy soil that for a compacted clay soil. However, a clay soil with a good stable structure may have a hydraulic conductivity higher than a sandy soil (Shalhevet 1965).

Several field and laboratory methods are available for evaluating the hydraulic conductivity of the soil. Complete descriptions of the various methods, with their advantages and limitations, are given in Klute (1986).

The two most commonly used techniques are (1) the auger-hole method for field determination of K under a water table, and (2) the undisturbed core sample method, for laboratory determination when there is no water table.

Evaluation of the mean effective hydraulic conductivity of a field to be drained is critical for proper drainage design. When it is possible to construct one or two short drainage lines in the field, prior to draining the entire field, or when a drainage system has already been constructed under similar conditions, the potential-discharge method provides the best means of determining K. This method utilizes drainage theory that relates the drain pipe discharge rate (q) to the hydraulic head midway between drains (h) as expressed in Eq. (5.11) below. When an impermeable layer exists at great depth, a plot of q vs h gives a straight line with slope K. With a shallow impermeable layer, a plot of q/h vs h will provide K above and below the drainage depth.

The drainable porosity of the soil is the pore volume that drains out when the water table is lowered a specified distance. It depends on soil texture and structure and will generally be higher for sandy and loamy soils and lower for a compacted clay soil. The drainable porosity affects the extent and rate of water table fluctuations as a result of water inputs from rain or irrigation. Under steady state conditions, when the water table position is in equilibrium with rainfall, the drainable porosity is of minor importance. Under nonsteady state conditions, the drainable porosity of the soil will determine the height of water table rise and the rate of its descent, and thus might have a profound influence on drain spacing.

The drainable porosity, p, may be approximately estimated from the relationship

$$p = \frac{Q}{m} t, \tag{5.10}$$

where Q = change in drain discharge in units of length per unit time and m = change in water table height above drain level midway between drains at time interval t. The assumption implicit in Eq. (5.10) is that the water table drops without change in shape. Bouwer and van Shilfgaarde (1963) suggested multiplying the equation by a factor 1/c, where c = 0.8–1.0, in order to correct for this discrepancy.

Determination of Drain Spacing. It is assumed in this section that a parallel system of buried drains is desirable, which is the most common drainage conditions in irrigated fields, especially in a situation when effluent water is used. Other situations may occur, such as seepage from unlined canals, artesian pressure and localized springs which will require relevant solutions (an interceptor drain, pumping, etc.).

Drain spacing is generally determined using a drainage equation, several of which are available, each applicable to a different situation. They all describe the quantitative relationship between the factors governing flow of water to drains. These factors are: (1) the permeability or hydraulic conductivity of the soil (K); (2) the hydraulic head (h) or the height of the water table above the drain level; (3) drain spacing (S), which together with the hydraulic head, determines the hydraulic gradient, i.e., the driving force for flow; (4) depth to an impermeable layer (D); and (5) the rate of water discharge from drains (q). In addition, under nonsteady state conditions, the drainable porosity (p) will also govern flow, as pointed out above.

Most drainage situations are nonsteady, i.e., flow in drains will vary considerably with time, due to variability in the rate of water input by rain or irrigation. Nevertheless, steady-state solutions for drainage design are generally adequate for most situations, and are much simpler to apply. The most frequently used solution is the one developed by Hooghoudt (Eq. 5.11), which is based on the assumption of horizontal flow to drains (the so-called Deupoit–Forchheimmer, or DF, assumption) with appropriate correction for flow convergence near the drains (Luthin 1965):

$$S^2 = \frac{8\,K_1dh}{q} + \frac{4K_2h^2}{q}. \tag{5.11}$$

The first term on the right-hand side of Eq. (5.11) represents the flow below the drain line level and the second term represents flow above this level. Consequently, in a layered soil, K_1 may be taken as the hydraulic conductivity of the layer below and K_2 of the layer above the drain level. When drains are laid on an impermeable layer ($K_1 = 0$), the first term may be neglected; when the impermeable layer is infinitely deep ($D \gg h$), the second term is negligibly small.

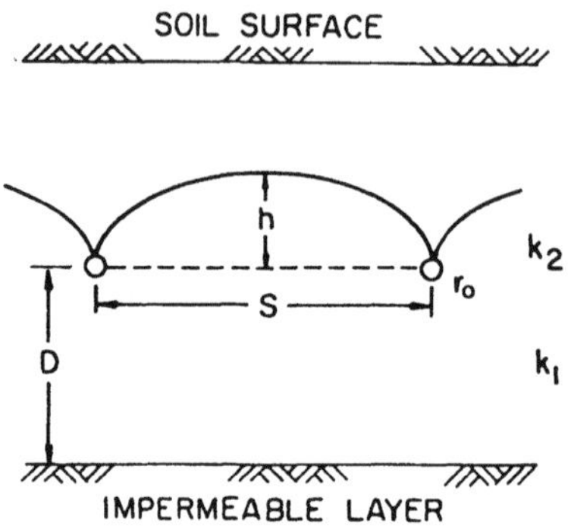

Fig. 5.6 Cross-section of a tile-drained field. The *symbols* are described in the text

Table 5.13 Equivalent depth to an impermeable layer, d, in m to be used in Hooghoudt's equation: $r_0 = 50$ mm, D = depth to impermeable layers, S = drain spacing

	S (m)					
	20	25	30	40	50	100
D (m)						
0.5	0.47	0.47	0.48	0.49	0.49	0.49
1.0	0.83	0.86	0.88	0.91	0.92	0.96
1.5	1.08	1.15	1.20	1.26	1.30	1.40
2.0	1.25	1.35	1.43	1.55	1.62	1.79
2.5	1.38	1.52	1.63	1.80	1.89	2.16
3.0	1.46	1.63	1.76	1.97	2.12	2.49
4.0	1.56	1.70	1.96	2.26	2.48	3.06
5.0	1.61	1.87	2.09	2.45	2.73	3.54
6.0			2.17	2.58	2.92	3.94
∞	1.62	1.94	2.24	2.84	3.41	6.08

In Eq. (5.11) d represents an equivalent depth to an impermeable layer and depends on drain radius r_0, drain spacing S, and the actual depth to an impermeable layer D (see Fig. 5.6). Values of d are given in tables or nomograms. Table 5.13 gives d values for various spacings and for $r_o = 50$ mm.

5.7 Monitoring

The use of sewage effluent for irrigation requires a broader monitoring program than that for natural waters (Bouwer and Idelovitch 1987). The objectives of the monitoring are to determine if the effluent quality requirements are being met, and if any corrective action is necessary to assure proper functioning of the entire system and to protect the environment.

Specific monitoring programs including sampling procedures and frequencies, as well as analytical methods, should be established for each component of the system. This includes the treatment plant, storage reservoirs, the irrigation water quality, the irrigated soils and crops and the receiving groundwater and surface water in the irrigated area. In some cases, especially when primary effluent or another odor-emitting wastewater is used, air pollution monitoring may also be necessary.

Most of the literature on monitoring programs for water quality are related to potable water quality. However, many of the general principles are applicable to irrigation with sewage effluent.

The specific site monitoring program should be based on in-depth preliminary studies, which should be conducted to determine ambient conditions of the site and water source and their stability, and on economic considerations (see also Sect. 5.1).

Monitoring programs of groundwater and surface water, are generally prescribed by the regulatory authorities. Groundwater quality can be monitored through well water analysis. Observation wells should be located in the irrigated area some distance both downstream and upstream from the relevant aquifer for comparison.

Monitoring the quality of the final effluent in the treatment plant, including indicator parameters such as BOD, TOC, suspended solids, pH and ammonia, is essential for the control of the treatment plant. Monitoring water quality in storage reservoirs at different depths is important to detect anaerobic conditions, algae and zooplankton development. Monthly or bimonthly measurements of clearness, temperature, soluble oxygen, pH, suspended solids, BOD and microscopic biological observations can help in determining the necessary depth of pumping of irrigation water, the management of the reservoir and the corrective measures taken to prevent adverse odors and the clogging in the irrigation system. Standard methods for such measurements are available (e.g., Taras et al. 1971).

Irrigation water quality assessment includes additional parameters as specified in Section 2.8. Since variations of such quality parameters as salinity, sodicity, toxicity and nutrient content may be larger in sewage effluent than in natural water sources, sampling and testing during the irrigation season should be more frequent. This is important in order to determine leaching requirements, or the necessity for blending with water of higher quality and for fertilizer management.

Annual or biannual soil sampling and testing may be sufficient to monitor changes in soil properties resulting from irrigation with sewage effluent. The sampling should include measurement of some physical properties such as bulk density or penetrometer resistance, indicative of soil compaction, observations of restrictive layers and measurements of aeration conditions. The electrical conductivity and SAR of the soil solution extracts should be determined at least before and after the irrigation season for proper irrigation management. It is also recommended that the level of trace elements of concern be monitored in the soil every couple of years to observe accumulation rate and avoid toxic levels. Soil monitoring requires the establishment of a sampling pattern, such as a marked grid to denote sampling points. Such a pattern should be based on the spatial variability of the specific site, but samples should always be composite of several subsamples around the sampling point to minimize influence of microvariations caused by plants, animals, fertilization and irrigation.

Plant tissue analysis is generally more revealing than soil analysis with regard to deficient or toxic levels of elements, or results of adverse management conditions. Thus, a regular plant tissue monitoring program, for short- and long-term effects, is strongly recommended for crops irrigated by sewage effluent. Such a program can be helpful in detecting deficiencies and toxicities before visual symptoms and damage occur, as well as warning of concentrations of elements which might be toxic to livestock or the public consuming the crops. A detailed account of soil and plant testing procedures is given by Walsh and Beaton (1973), Page et al. (1982) and Reisenauer (1978). This information can be used for fertilization recommendations under specific field conditions.

Finally, a bacteriological monitoring program (e.g., fecal coli) of crops to be marketed for human consumption should be included.

Chapter 6 Irrigation Systems for Sewage Effluent

6.1 Considerations in Irrigation System Design

6.1.1 Introduction

The primary objective of irrigation with reclaimed wastewater is, in most cases, to maximize crop production, disposal being a secondary benefit. By and large, any irrigation method used with water from conventional sources may be adapted, with only minor modification, for use with sewage effluent.

Irrigation methods may be classified into two general types: (1) surface or gravity irrigation (furrow and border irrigation) and (2) pressure irrigation (spray and drip irrigation). Both types are used for sewage effluent irrigation. The selection of the irrigation method is based on a variety of criteria, including economic considerations, land topography and soil physical characteristics, type of crops, availability of skilled labor, water quality and farming traditions. After selecting the irrigation method, a detailed design of the system's components will be necessary, for which consultation with a competent irrigation engineer is advisable. For further details on irrigation engineering the reader is referred to Jensen (1980), Finkel (1982) and Benami and Ofen (1983).

The purpose of this section is to present the basic principles of those systems that are applicable for sewage effluent irrigation, their limitations and advantages.

6.1.2 Theoretical Considerations

The net irrigation requirement, NR, in mm, is calculated from

$$NR = (E_c - P)/(1 - LR), \tag{6.1}$$

where E_c is the crop water consumptive use requirement, P is precipitation, and LR is the leaching requirement. The NR and E_c may be calculated on a monthly or weekly basis (see Sects. 3.1 and 5.2).

Once the type of irrigation system is selected, and its application efficiency is estimated the gross water requirement (GR), in mm is calculated from

$$GR = NR/e_a, \tag{6.2}$$

where e_a is the efficiency factor. e_a defines the fraction of the total amount of water applied to the field that actually enters into the soil and stays within the active root zone. e_a varies between 0.75 and 0.85 in most sprinkler systems, and is somewhat higher in a well-designed drip irrigation system. It is 0.6–0.8 or even lower in most surface irrigation systems, although in precise leveled border systems it may be as high as 0.90.

If the percolating water from an effluent-irrigated area flows to a potable groundwater aquifer or to a surface stream, restrictions are usually placed on the composition of the deep percolating solution, e.g., the allowable nitrate N concentration in the percolating water according to EPA (1981) is 10 mg/l. The allowable annual irrigation system capacity Q_N, in mm, based on nitrogen limitation is then calculated from

$$Q_N = \frac{10^2\, U_N - C_P^N (E_c - P)}{(1 - f)\, C_N - C_P^N}, \tag{6.3}$$

where U_N is the net nitrogen uptake by the crop, in kg N ha^{-1} yr^{-1}, C_P^N is the allowable N concentration in the percolate, in mg N/l, C_N is N concentration in the applied effluent, in mg N/l, and f is the fraction of the applied N lost by soil processes. E_c and P are the yearly crop water consumption and precipitation, respectively.

If Q_N is less than the annual gross water requirement, GR, and there is no way of ensuring sufficient mixing and dilution of the percolate with the flowing water in the aquifer or stream, then either a different crop, with higher N uptake, should be selected or the quality of the effluent should be upgraded in the sewage treatment plant. Similar calculations can be made for other hazardous constituents of the effluent; e.g. Cd, trace organics.

The frequency of irrigation and depth of water applied per irrigation are interrelated and serve as management parameters (see Sect. 5.2). The relationship between the irrigation interval, t_0, in days, and the depth of water applied per irrigation, D_i, in mm, used for the system design is given by

$$D_i = t_0 \times E_d / e_a, \tag{6.4}$$

where E_d is the mean daily ET rate of the crop, in mm/day, and e_a is the efficiency coefficient.

Application rate is greatly dependent on soil properties, infiltration rate and soil permeability being the dominant factors. Infiltration rate, IR, decreases with time as water application continues (Fig. 5.1). There is evidence that this decrease is greater with wastewater. The IR is lower when the soil is wet, but it may be restored following a drying period. Tests to determine IR should be compatible with the irrigation method to be used and should employ water of similar composition. Day et al. (1972) reported reduction in infiltration rate after several years of wastewater application of rates equivalent to normal irrigation practices. This effect would be expected to appear earlier at high rates of application. A pilot study is recommended, especially when a large project using surface irrigation methods is considered. Many effluents have high BOD, thus proper aeration of the soil is very important. Soil permeability is the major factor determining the time required for the soil to reaerate after irrigation (see Sect. 5.6). With effluent of high BOD, the interval between irrigations should be long enough to allow the soil to dry, or the management allowable soil water deficit (MAD) should be larger than with regular supply water. The soil infiltration rate should be maintained by appropriate management practices.

The system discharge (or capacity) Q in m^3/h is expressed as:

$$Q = \frac{10\,D_i\,A}{t\,t_0}, \tag{6.5}$$

where A is the area to be irrigated in ha, t is operating time, in h/day, t_0 is the irrigation interval in days, and D_i is depth of water per irrigation, in mm, (D_i/t is the application rate). Frequently, in the case of sewage effluent, Q is fixed. Hence, for a certain crop and soil the area A is computed from [Eq. (6.5)]. Where effluent is supplied from a pond or reservoir, Eq. (6.5) can be used to determine the required pump characteristics.

6.1.3 Surface Irrigation Systems

The basic principle of surface irrigation is that water is applied on the ground at one end of the field and is allowed to flow by gravity to the other end. As it advances, the water infiltrates into the soil and spreads over the field. The most widely used gravity irrigation methods are furrow and border. In border systems, the water is taken out from a ditch, in which the water level is usually higher than the ground surface.

Graded borders are leveled land strips (usually 10–20 m wide), of a small, uniform longitudinal slope. Border systems are generally employed on extended areas of relatively flat topography for irrigation of grain, forage and field crops and are thus suitable for sewage effluent use. Specially designed border systems can also be used on hilly topography, where the supply ditch runs down the main slope and feeds water to border strips which are laid just slightly off the contour line. Border systems are practical where the available water discharge is high ($>100\,m^3/h$).

Recently, the level basin technique has been widely introduced, especially in southern U.S.A. In this system, the basin is leveled in all directions and much of the infiltration into the soil occurs after the water advance has been completed. High flow rates and/or low infiltration rates enable the use of large basins. These systems are easily managed and have a much lower labor demand than borders or furrow systems. (Dedrick et al. 1982).

Furrow irrigation, which is suitable in cases of small water discharge, consists of shallow, narrow, gently sloping furrows and water is usually applied by gated, aluminum pipe. As it advances in the furrow, the water infiltrates into the soil profile and spreads laterally to provide the desired wetting of the soil. The crops are grown on raised beds along the furrows. Furrow irrigation is well suited for row crops irrigated with sewage effluent.

Usually, the initial capital investment in surface irrigation systems is much smaller than that needed for pressure-pipe systems. But, although modern design methods have improved the accuracy and efficiency of water application, good irrigation requires skilled and experienced labor.

Most surface irrigation systems produce runoff (also referred to as tail water). When effluent is used there should be no runoff or tail water from the irrigated area except to an approved disposal site, from which the water can be returned to the effluent pond, or reservoir, or recycled into the irrigation system. Surface irrigation methods are in general not amenable to mchanized and automated farming, although automatic, remotely controlled gates are available. Periodic maintenance of the leveled surface is required to maintain uniform grades to avoid puddling, particularly when irrigating with wastewater.

The design of surface irrigation systems is largely based on experience and direct field studies. It is important to estimate seepage losses from open, unlined channels, in order to assess conveyance losses and compute the required flow capacity of the delivery system, as well as to evaluate seepage into the groundwater.

Good drainage is most important; a rising water table due to overirrigation and ditch seepage, which occurs frequently with surface systems, presents a hazard due to both a restricted root zone and to increased salinity. Thus, it is inadvisable to apply these methods to shallow soils or where a high water table exists, especially if the treated effluent is of poor quality.

Surface irrigation methods are suitable for soils of medium to fine texture where a satisfactory application efficiency can be attained. On highly permeable and variable soils, uniform water distribution cannot be achieved. Very fine textured soils of very low infiltration rates are also not suitable for surface irrigation. Since in most surface irrigation systems, water flow rates are generally quite high, these methods should be avoided on easily eroding soils. This is particularly important when using effluent that may contaminate the upper soil layer.

Furrow irrigation may cause accumulation of salts along the longitudinal strips midway between the furrows, where the water and salts move upward toward the soil surface (Ayers and Westcot 1985). When water salinity is high, special management practices for planting or seeding and shaping of the planting beds are required to maintain high production levels.

A very important advantage of furrow irrigation methods for use with sewage effluent is that the plant foliage remains dry, which also prevents leaf burn when the salinity of the water is high. Surface irrigation is unaffected by wind conditions; thus there is no danger from wind-drifted contaminated aerosols.

For more detailed information on the various surface irrigation methods the reader is referred to Booker (1974).

6.1.4 Pressurized Irrigation Systems

Sprinkler Irrigation. Sprinkler systems are basically rainfall simulators. In the sprinkler the pressure head of the water is converted into a velocity head. The jet ejected from the nozzle breaks into small droplets, which fall over an area the size of which depends mainly on the operating pressure and type of sprinkler. Sprinklers vary from low flow-rate sprinklers placed at desired spacings along a lateral to giant or gun sprinklers mounted on carriages, which facilitate rapid coverage of large areas.

The conventional sprinkler system is semiportable. One or more laterals irrigate on both sides of a single, permanently placed mainline pipe equipped with take-off valves. The laterals are stationary while irrigating and are moved to new positions along the main line until the entire field has been covered. These systems are suitable where crop-soil-weather conditions permit a frequency of irrigation of at least a few days.

Rotating impact sprinklers, with single or double nozzles, are the most commonly used of all commercially available sprinklers. These sprinklers are suitable to a wide range of operating pressures, discharges and application rates and practically all crops and soils. Spacing of sprinklers on the laterals and spacing of the take-off valves on the main line are designed so that the water distribution patterns from the sprinklers give nearly complete overlap.

For sewage effluent irrigation, moving of the laterals by hand is not advisable, because it involves a great deal of direct physical contact of the irrigator with the effluent. Use of mechanically moved systems (side-roll, tow-move), in which the lateral is mounted on wheels or skids, will reduce labor and health hazards.

Center-pivot systems include a lateral pipeline, generally made of galvanized steel, anchored to a fixed central pivot structure around which it continuously rotates covering a circular area. Water is supplied to the lateral pipe through the pivot. The lateral pipe with the desired sprinklers is supported on 8–10 self-propelled towers on wheels at about 30–60 m apart. The sprinklers are arranged on the lateral so that they can produce uniform water distribution while making a complete revolution, typically every 60 h. The sprinkler can be replaced by spray nozzles on vertically suspended flexible tubes close to the crop to reduce aerosol drift. The system can be fully automated. The smallest practical unit for agricultural crop irrigation is about 4 ha.

Traveling lateral or side-roll systems that move continuously in a rectilinear fashion are available and becoming increasingly popular for row crops (Fig. 6.1). In these systems sprinklers can be replaced by trickle emitters that apply the water on the ground, thus eliminating aerosol drift. In these systems the lateral is fed either from a flexible hoze or by a traveling pump unit that pumps the water from a ditch.

The self-propelled *gun-sprinkler traveler* consists of a giant sprinkler (with discharges ranging from 20–200 m^3/h at operating pressures of 40–70 m) mounted on a trailer to which a flexible hose (75–130 mm diameter and up to 400 m long) is connected. The unit is pulled at a predetermined fixed speed, along a path from one end of the field to the other, by a cable wrapping on a rotating reel. A traveler can irrigate unattended a strip as long as 800 m. The traveling gun system is adaptable to odd-sized and -shaped fields with a wide range of topography and soil conditions. Gun sprinklers have a high production of aerosols, hence, they may be used in remote fields or where a large buffer zone can be accommodated for sewage effluent irrigation.

Fig. 6.1 A 375-m-long linear-move irrigation system used to irrigate cotton with sewage effluent

Solid-set or permanent sprinkler systems are the most adaptable for projects of effluent disposal at high frequency and application rates. These systems may also be equipped with gun sprinklers. Solid-set systems are well suited to complete automation but can also be operated manually. They are adaptable to a wide variety of terrains and can be the only suitable system for sprinkler irrigation of tree crops. The high initial investment in such a system is compensated for by the reduced labor requirement and increased versatility and efficiency.

An effluent sprinkler irrigation system should be designed to minimize aerosols and windblown spray, which represent a potential health hazard. It is recommended to select low sprinklers with low-pressure nozzles. In mechanically moved systems the lateral can be fitted with spray nozzles directed downward and applying the effluent close to the ground. It is also recommended to irrigate during periods of low wind velocity and during hours when people are least expected in the vicinity of the irrigated fields. Windbreakers and/or buffer zones should be allocated around the site. It is also advisable to use effluents that have been effectively disinfected beforehand. During hot weather, sprinkler irrigation, particularly when using high-salinity water, may cause leaf burn. Hence irrigation at night, when evaporation is low, is preferable.

Where possible, the system should be flushed with fresh water after each use to minimize corrosion hazard, clogging of sprinkler orifices and hazard of formation of septic conditions in the system.

The design of a sprinkler irrigation system for wastewater application is similar to the design for regular water systems. Special equipment may be required only if the wastewater is abrasive or corrosive or has a high solids content.

Drip Irrigation Systems. Drip irrigation may be described as a method of frequent, slow application of water through a network of many small emitters placed on, or very near to, the ground surfaces. Drip irrigation systems are generally of the solid-set type, operating under low pressure heads. In the emitter the pressure head is dissipated by means of orifices, vortexes and a tortuous or long flow path. The narrow passages in the emitter makes it vulnerable to clogging by suspended particles. Therefore, drip irrigation requires filtered water.

The water trickling from the emitter flows through the soil by capillarity and gravity. The size of the wetted area depends mainly on the emitter discharge rate and soil type and, due to the high frequency of water application, tends to be steady. The ease with which water can be applied using automation makes it possible to irrigate with high efficiency sands, shallow soils and hill sides.

Drip irrigation has certain specific advantages with regard to the use of treated sewage effluent. Weed growth, which may be a problem because of the high nutrient content of the effluent, is restricted due to the partial wetting of the ground. The dry soil strips between laterals permit uninterrupted cultivation, which is particularly important in orchards and vineyards. Due to the slow water application and the partial wetting of the soil, runoff is restricted and deep percolation can be better controlled. The better aeration regime in the root-zone achieved by drip irrigation may be particularly advantageous when using effluent with high BOD. Foliage and fruit are maintained dry during irrigation, thereby the hazard of aerosol contaminants deposition is eliminated. Drip irrigation is highly suitable for rapid, efficient application of fertilizers and chemicals (chemigation). Finally, drip irrigation is suited for all types of soil and topography, requiring no special land-surface preparations, and automation can be easily incorporated into the system.

Drip irrigation systems are generally initially expensive. With its supporting equipment, a drip irrigation system generally costs more than a center-pivot system, but less than a solid-set sprinkler system. A drip irrigation system would not be economical where large amounts of pipe and emitters per land unit are required, as would be in the case of forage and cereals irrigation. Approximately 10 000 m of pipe and 10 000 emitters are needed to irrigate a 1-ha plot with lateral spacing of 1 m and emitter spacing of 1 m along the lateral. The annual operation costs of a drip irrigation system is generally considered to be comparable with other irrigation methods. It was, however, found that the fixed costs per unit land area of cotton in the USA was 2.3 times higher for a drip irrigation system than for a conventional furrow irrigation system (Nakayama and Bucks 1986). Actual costs may vary greatly depending on type of crop, soil, the selected system components, filtration equipment and level of automation.

Drip irrigation may be advantageous in irrigating small, irregular plots, or of "marginal soils", such as steep slopes, rocky hillsides, road embankments and highly permeable, low-waterholding capacity soils.

Drip irrigation is applicable where the total available effluent discharge is small. For example, the discharge required to irrigate a hectare of cotton by a drip-system (2 × 1 m spacing of 4 l/h emitter) is about 20 m^3/h and the operating pressure head is 5–15 m. For a conventional sprinkler system the required discharge is about 80 m^3/h and the pressure head 25–35 m. Approximately 200 m^3/h would be required in case of a furrow irrigation system of 4 m^3/h per furrow, and for border strip irrigation the required discharge may reach 1000 m^3/h.

Emitter clogging is a major problem associated with drip irrigation, particularly when using effluent. Clogging of even a small percentage of the emitters can severely affect the uniformity of water application. Obviously, the larger the passages in the emitter, the smaller the clogging hazard. Hence, for effluent irrigation it is advisable to design the system for use of emitters of high discharge rate. Clogging by more or less rigid, large particles can be controlled by proper filtration. A filtration system, with the capacity to deliver the designed water flow and the ability to remove particles of sizes larger than about half the narrowest passage of the emitter, are mandatory, regardless of the irrigation water source (see Sect 6.7). Periodic flushing of laterals has also been found to be beneficial in reducing clogging. However, even with the finest filtering system, clogging can arise from fine particles passing the filter that are combined by bacterial slime and cemented by precipitated $CaCO_3$. The nutritional nature of effluent and absence of light in the pipe system may cause proliferation of certain microbial filaments. These can clog not only emitters but also the lateral pipes and the filtering system. The most effective preventive measure for these types of clogging is chlorination (see Sect. 6.4).

Detailed information on drip irrigation system design and operation can be found in Jensen (1980), Benami and Ofen (1983), Dasberg and Bresler (1985) and Nakayama and Bucks (1986).

Recent developments in traveling trickle irrigation technology has contributed to the solution of the clogging problem and has reduced the need for extensive pipe network (Phene et al. 1985). Such systems are especially suitable for irrigation with sewage effluent.

6.2 Piping Systems

6.2.1 Materials

Pipes are major components of most irrigation systems. Aluminum, polyvinyl chloride (PVC), polyethylene (PE), steel, concrete and asbestos cement (AC) are the most common materials used for irrigation system pipelines. Although most secondary effluents do not cause corrosion, awareness of this possible hazard is most important when wastewater is used.

Almost all above-ground pipelines of large sprinkler irrigation systems are aluminum. The main advantages of aluminum pipes are their light weight, their resistance to mechanical damage and to sunlight and the ease and rapidity with which they can be assembled and dismantled. They are, however, vulnerable to damage under freezing conditions (pipe splitting) when they are less than half full, as may occur at low points after draining the pipeline. Aluminum pipes may be attacked by fertilizers and chemicals applied with the water and must be flushed thoroughly after application.

In drip irrigation and some sprinkler systems, black polyethylene pipes are used as laterals. This pipe is generally inexpensive, is unaffected by chemicals in the water, is resistant to sunlight and has a relatively long life. Polyethylene pipes are easily damaged by rodents and woodpeckers.

Generally, PVC, AC and steel pipes are used for mains and submains. PVC is probably the most suitable pipe for wastewater transportation as it is resistant to corrosion and chemical attack. A PVC pipe has less internal friction than other pipelines and is durable when protected from surge pressures. Since PVC will deteriorate in sunlight, it is advisable to bury PVC pipelines. This will also help protect them from damage by farm machinery, animals, etc.

Asbestos cement pipes are also durable and corrosion resistant (except for high levels of sulfate ions) and are known to be suitable for most waste water systems. However, they are liable to cracking when not appropriately supported.

Steel and concrete pipes are the least suitable for wastewater systems. Although they are strong and durable, they have a tendency to corrode on both the inside and outside. Calcareous scale deposits formed with hard water act as a protective coating. Soft water containing little calcium is, as a rule, more corrosive. Chloride ions are also highly corrosive, especially towards steel and some stainless steels, as they tend to break the protective film and initiate corrosion. Yellow brasses ($>15\%$ zinc) are very susceptible to dezincification by water with a high chloride content. Red brass and copper are not as easily affected and are preferable in sprinklers, drippers and fittings when wastewaters are used. Water or soil with a high level of dissolved salts will be more conductive than those of low salinity and are thus more corrosive. Contact points of dissimilar metals, found in sprinkler or in other devices in the irrigation system, are especially vulnerable to salinity corrosion.

Concrete and cement deteriorate in the presence of high levels of sulfates. Under anaerobic conditions or low oxygen concentration in the presence of organic matter, H_2S will generally be formed and produce sulfuric acid. Such conditions are readily developed in soils, storage basins and pipes with effluent of high oxygen demand. Steel and copper are also attacked by sulfuric acid. Concrete corrosion can occur also with water containing $>15\,mg\ NH_4/l$. Ammonia is also very corrosive towards copper.

Most problems occurring with water transportation and distribution systems can be prevented through proper design. In addition to selecting a pipe size to fit the pipeline length, discharge rate, and allowed friction loss, a good design includes choosing appropriate pipe material, pressure class and proper location of blowoffs and air release valves. The latter are essential for satisfactory operation of the irrigation system, particularly under varying topography.

6.2.2 Safety

An important consideration in the design of wastewater irrigation projects is absolute physical separation of wastewater transportation and distribution pipelines and appurtenances from the potable water system. Separation between the two water pipelines should be maximized in order to minimize accidents during construction and maintenance operations breaks and possible leakage of effluent into domestic water lines, and to prevent accidental cross-connection. Effluent pipelines should be properly identified by color coding. All valves and outlets from effluent systems should be tagged with an appropriate warning that the water is unsafe for drinking, in addition to being color-coded. Posting of warning signs on effluent-irrigated areas and storage ponds is also advisable. Records of all plans of effluent pipelines should be easily available.

6.3 Storage and Pumping

6.3.1 Storage

A storage facility is, in most cases, a crucial link between the wastewater treatment plant and the irrigation system, and facilitates irrigation scheduling. The advantages of having a storage pond or reservoir as part of the irrigation system are:

1. Equalization variations in effluent flow from the treatment plant, and storage of excess effluent when flow exceeds irrigation demand (during the rainy seasons), especially if no alternative way of disposal exists
2. Meeting peak irrigation demands in excess of effluent flow
3. Minimization of disruptions in irrigation during period when effluent is not delivered from the treatment plant, if no alternative water source is available.
4. A safeguard against the possibility of unsuitable effluent entering the irrigation system. Storage provides a possibility for additional treatment and dilution.

The main disadvantage is economic. Reservoir capital costs and operation and maintenance costs depend on storage volume and can be considerable. Total irrigation water requirements on a monthly or weekly basis are needed for storage-volume calculations. Reservoirs are usually constructed with small earth dams (USBR 1974) but lining and embankment protection may be required.

6.3.2 Pumps

Most irrigation systems require some sort of pumping to lift water from ponds, reservoirs or canals and to provide the pressure head needed for the specific irrigation system.

Horizontal centrifugal pumps are the most common in irrigation systems. These pumps are usually connected directly to electric motors and can be purchased as a complete unit, requiring only a hookup to the intake suction and the discharge pipes. The pumping unit is positioned above the water surface close enough to it for ease of priming. It is important to avoid submergence of electric motors.

The power (HP; in horsepower units) requirement of a pump, often called the brake horsepower, is given by

$$HP = Q \cdot H/273\,\eta, \tag{6.6}$$

where Q is the discharge, or capacity (in m^3/h); H is the total pressure head, including friction head loss (in m); and η is the pump efficiency (the overall efficiency of a new electric motor and pump should be about 0.65 or 65%). The relation of the change in impeller speed, n in r.p.m., or in the impeller diameter, D to the change in pump discharge and power requirements is given by

$$\frac{Q_1}{Q_2} = \frac{n_1}{n_2}; \quad \frac{H_1}{H_2} = \left(\frac{n_1}{n_2}\right)^2; \quad \frac{HP_1}{HP_2} = \left(\frac{n_1}{n_2}\right)^3; \quad \frac{n_1}{n_2} = \frac{D_1}{D_2}. \tag{6.7}$$

Pumps have well defined operational characteristics. The selection of a pump for any particular operating conditions of the irrigation system is made by using the pump characteristic curves, provided by the manufacturer. The characteristic curves present the interdependence between the capacity, head and efficiency at a given speed (Fig. 6.2). Evidently, the pump will operate over a wide range of head-capacity conditions, but it will operate at high efficiency only under a limited range of conditions around the peak of the efficiency curve (working point in Fig. 6.2). The shape of the H–Q curve is primarily dependent on the specific speed of the impeller, but is related also to the impeller type, the form of the casing and the energy recuperators employed.

There are three basic types of impellers in centrifugal pumps: open, semi-enclosed and enclosed. The enclosed impeller has the highest efficiency, longest life and lowest maintenance, but it has the highest cost, and is prone to clogging

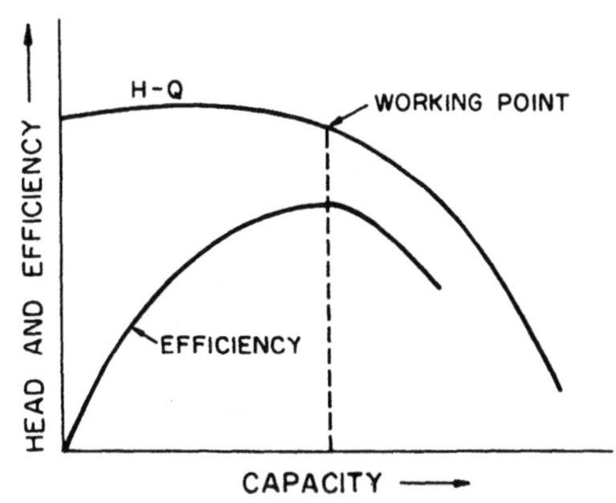

Fig. 6.2 Characteristic curve for centrifugal pumps. The pressure head (*H*) – discharge capacity (*Q*) curve, and the working point located at the peak of the efficiency curve

by suspended matter. These pumps can be safely used with secondary treated effluent. For primary effluent the semienclosed impeller, of lower efficiency and lower cost, is a better choice. The open impeller, with the lowest efficiency and lowest cost, is practically nonclogging and is adapted for sludge application systems.

Sediments in the water will cause wear of any pump and may require periodic replacement of impellers and volute cases. It is advisable that a screen filter be installed at the intake suction side of the pump.

Axial-flow (propeller) pumps are a better choice for high-discharge, low-head duties, which are typical for surface irrigation systems. Their input power demand is less for low-head pumping than that of centrifugal pumps, and their initial cost is relatively low. Axial pumps are commonly vertically mounted directly over a sump. Submergence depth should follow the manufacturer's recommendations.

Pumps can be connected in series to increase the pressure head for a given discharge or in parallel to increase discharge for a given pressure head. Pumps for irrigation systems may be powered by any power source. Beside electric motors, diesel and gasolene engines are also common, and for small irrigation systems, use of a farm tractor take-off unit may reduce initial investment. Electric motors are most easily automated and generally require very little maintenance. It is very important to obtain correct matching of pump, motor and drive for high efficiency pumping.

6.4 Chlorination

Inclusion of chlorination as part of the effluent irrigation system is common practice, intended mainly to reduce health hazards. Chlorination also helps to prevent growth of bacterial slimes and algae within the transfer and distribution system (sprinkler head, emitter orifice or supply line), thus reducing clogging hazards, particularly in drip irrigation systems. Addition of relatively small doses of chlorine can effectively reduce H_2S and other offensive odors of effluent. Chlorination is highly recommended for spray irrigation systems, especially where effluent drift may reach the public, and is essential in drip systems to overcome clogging hazards.

The economic feasibility of treating large volumes of water is the principal reason that chlorine gas (Cl_2), sodium hypochlorite (liquid NaOCl) and, to a lesser extent, calcium hypochlorite [solid or powdered $Ca(OCl)_2$] are employed almost exclusively in irrigation systems. The chemistry and application principles of chlorination for irrigation waters are similar to those used in municipal drinking water and wastewater treatment facilities or in swimming pools. Chlorination of water involves the following chemical reactions: chlorine (Cl_2) hydrolyzes rapidly in water, forming hypochlorous and hydrochloric acids:

$$Cl_2 + H_2O = HOCl + H^+ + Cl^- \qquad (6.8)$$

$K_d = 3.94 \times 10^{-4}$ at 25 °C.

At Ph > 3 and chloride concentrations < 1000 mg/l, free chlorine molecules are virtually absent in solution. The hypochlorous acid is a weak acid and partially dissociates into a hydrogen ion and a hypochlorite ion:

$$HOCl = OCl^- + H^+ \tag{6.9}$$

$K_d = 2.9 \times 10^{-8}$ at 25 °C.

The relative amounts of OCl^- and HOCl ("free available chlorine") in solution are pH dependent (Fig. 6.3). The percentage of HOCl, which is a stronger disinfectant than the hypochlorite ion, can be calculated from Eq. (6.9) as:

$$\%[HOCl] = 100/(1 + K_d/[H^+]) \tag{6.10}$$

The hypochlorite sodium and calcium salts dissociate completely in dilute aqueous solution to yield the hypochlorite ion:

$$NaOCl = Na^+ + OCl^- \tag{6.11}$$

$$Ca(OCl)_2 = Ca^{2+} + 2\,OCl^- \tag{6.12}$$

The hypochlorite ion subsequently reacts with a proton to give hypochlorous acid as indicated in Eq. (6.9). Thus, the hypochlorite salt hydrolysis is usually accompanied by an increase in pH. To enhance bactericidal effectiveness, acid may be added with hypochlorites to increase the percentage of hypochlorous acid. Chlorination is relatively ineffective above pH 7.5.

Chlorine, HOCl and OCl^- are very active and strong oxidizing agents and react with many chemicals found in wastewater. Ammonia, which is commonly present in effluent and in swimming pools, reacts with chlorine to form mono-, di- and trichloramines ("combined available chlorine"):

$$NH_3 + HOCl = H_2O + NH_2Cl \text{ (monochloramine),} \tag{6.13}$$

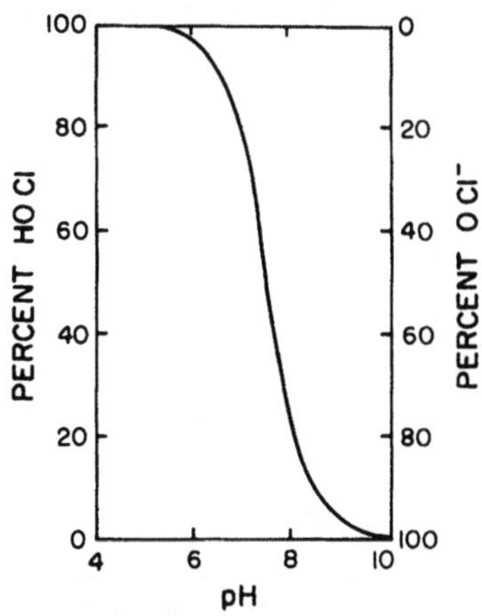

Fig. 6.3 Effect of pH on relative concentrations of $HOCl^-$ and OCl^- in aqueous solution (at 25 °C, K_a-$[H^+][OCl^-]/[HOCl] = 2.9 \times 10^{-8}$)

$$NH_2Cl + HOCl = H_2O + NHCl_2 \text{ (dichloramine)}, \quad (6.14)$$

$$NHCl_2 + HOCl = H_2O + NCl_3 \text{ (trichloramine)}. \quad (6.15)$$

The resulting chloramine forms and their relative amounts depend on pH, temperature and concentration of ammonia and hypochlorous acid. The most predominant species are the mono- and dichloramines. The off-taste and odor in swimming pools and drinking water are due to chloramines and not necessarily to chlorine. Trichloramine, the principal eye irritant in chlorinated water, is not stable at the pH of most natural waters. Many substances, such as Fe^{2+}, Mn^{2+}, H_2S and other reduced forms present in water, are oxidized very rapidly and convert the chlorine to the ineffective chloride. It is thus important to add sufficient amounts of chlorine to produce enough residual chlorine for controlling the microorganisms. Various chlorine test units are available. Portable units, in which the testwater is treated with an indicator crystal or solution and the resultant color compared on a color chart, disk or standard reference solution are suitable for field operations. Swimming pool test kits are usually based on orthotolidine indicator and give free and combined chlorine residual. Free residual chlorine can be measured by kits with the DPD (*N*, *N*-diethyl-p-phenylene diamine) indicator.

The addition of increasing amounts of chlorine to water followed by measurement of residual chlorine (combined and free) produces a curve commonly referred to as the breakpoint curve (Fig. 6.4). After meeting the initial demand of readily reactive species (the area between points 1 and 2 in Fig. 6.4) chloramines are formed between points 2 and 3, the relative amounts of each form being a function of pH. Oxidation of chloramines between points 3 and 4 results in a decline in the residual chlorine. Point 4 is termed the breakpoint, and addition of chlorine beyond this amount yields an increasing residual of free available chlorine. The reaction of chlorine with ammonia (optimum between pH 6.5 and 8.5) is usually completed in 30 min. The maintenance of residual free chlorine (commonly between 1 and 5 mg/l) minimizes the growth of microorganisms and slimes and causes inactivation of pathogenic bacteria. Chloramines, being less reactive than free chlorine, persist for longer periods after chlorination and can control bacteria and algae aftergrowth.

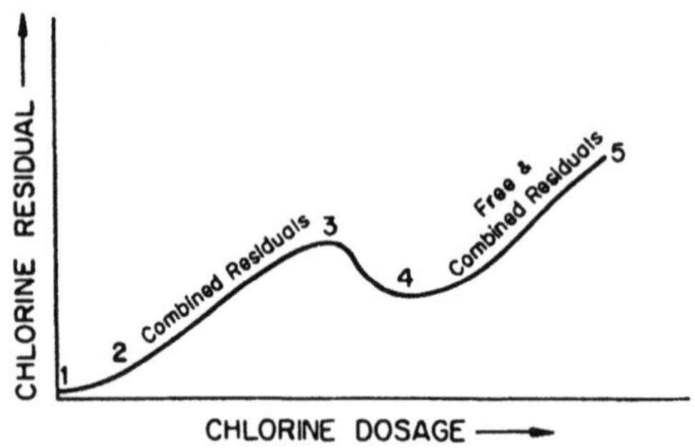

Fig. 6.4 Generalized breakpoint chlorination curve. The curve indicates the following changes in chlorine residuals: initial demand fulfilled (points *1–2*) formation of chloramines (points *2–3*), oxidation of chloramines (points *3–4*); addition of chlorine beyond point *4* results in increasing residual of free available chlorine

The amount of chlorine added is determined empirically, based on the desired residual and effluent quality. In the water disinfection process, the rate of kill is described by Chick's law (Chick 1908), stating that for a given concentration of disinfectant, the longer the contact time, the greater the kill. For continuous disinfection of irrigation water, a free residual chlorine concentration of 1–2 mg/l at the end of the distribution lines is generally recommended. When the aim of chlorination is only to reduce clogging hazard, particularly important in drip irrigation systems, intermittent treatments with 10–20 mg/l free chlorine and a contact time of 10–30 min has also proved to be satisfactory. The frequency of intermittent chlorination will depend upon the water quality and irrigation system characteristics. Excessive free residual chlorine in effluent, sprinkler-applied on plant foliage, can cause yellowing of leaves. However, such high concentrations will rarely be encountered in effluent irrigation. High chlorine residuals dissipate rapidly in the air. Residual chlorine of > 0.5 mg/l may affect plant foliage (Bower 1982). Greater experience is needed before more definitive recommendations can be given.

Chlorine gas is supplied as a liquified gas in cylinders under high pressure and may be dissolved directly into the main water supply line with the use of a metering device called a chlorinator. Chlorine gas greatly irritates mucous membranes, is toxic and very corrosive, and should thus be handled with appropriate precaution. Gas chlorination is widely used with systems that require large volumes of water, because it is generally the least expensive method.

Sodium hypochlorite solutions of 5–16% by weight of available chlorine (known also as "laundry bleach") are conveniently used for smaller irrigation systems. High concentration solutions are relatively unstable and are affected by light and heat and should therefore be stored in a sheltered area in corrosion-resistant tanks. It should be mentioned that use of sodium hypochlorite, will increase the level of sodium in the irrigation water and thus enhance soil sodicity hazards. Calcium hypochlorite, in solid form or in powdered form as chloride of lime (bleaching powder) and which must be dissolved in water, can also be used in irrigation systems. These compounds are similar to NaOCl with respect to properties and application. Sodium and calcium hypochlorites will react with emulsifiers, fertilizers, herbicides and insecticides and destroy their effectiveness, and should thus be applied separately. It should be noted that sodium and calcium hypochlorites have an alkaline reaction with water and that an acid may be needed to reduce the pH. A choice between different commercial products is usually made largely on the basis of cost per unit available chlorine, handling properties, availability, and also the initial cost of the injection system.

6.5 Precipitation of $CaCO_3$

Precipitation of calcium carbonate is common in water rich in calcium and bicarbonate and of high pH. The scale deposited in the irrigation system (in pipes, filters, regulating and metering devices, orifices, emitters and low-volume sprinklers) as a result of the use of such water can cause plugging and malfunction of the system.

The Langelier saturation index (Langelier 1936) that was developed for predicting carbonate scale deposition in boilers and later applied to sodium hazard (see Sect. 3.1) in soils (Bower et al. 1965) is useful for evaluating effluent quality with regard to scaling. The Langelier saturation index (LSI) is based on calculating the pH which a given water would assume when equilibrated with solid $CaCO_3$ (pH_c) and measuring the actual pH of that water (pH_a). The LSI is defined as pH_a–pH_c. Positive values ($pH_a > pH_c$) indicate that the water is supersaturated with respect to $CaCO_3$ and thus has a tendency for precipitating $CaCO_3$, whereas negative values ($pH_a < pH_c$) indicate the possibility of the water dissolving $CaCO_3$ in the system.

The following simplified computational procedure for evaluating pH_c, adopted from Nakayama and Bucks (1985), involves the use of the solubility product of $CaCO_3$ (K_{SP}) and the dissociation constant of HCO_3^- (K_d), defined as

$$K_{SP} = (Ca^{2+})(CO_3^{2-}), \tag{6.16}$$

$$K_d = \frac{(H^+)(CO_3^{2-})}{(HCO_3^-)}, \tag{6.17}$$

where parentheses designates the component activity.

Dividing Eq. (6.16) by Eq. (6.17) and redefining activities in terms of concentration by using activity coefficients γ gives

$$K_{SP}K_d = [Ca^{2+}][HCO_3^{2-}]\gamma_{Ca}\gamma_{HCO_3}/(H^+) \tag{6.18}$$

where brackets denote concentration. By expressing Eq. (6.18) in logarithmic form, using the conventional designation of the negative logarithm p and rearranging, pH_c can be evaluated from the following.

$$pH_c = (pK_d - pK_{SP}) + p[HCO_3^-] + p[Ca^{2+}] + p(\gamma_{Ca}\gamma_{HCO_3}). \tag{6.19}$$

To obtain $p[HCO_3^-]$ and $p[Ca^{2+}]$ from water analysis concentrations of HCO_3^- and Ca^{2+}, in mmol(charge)/l the following equations must be used

$$p[HCO_3^-] = 3.00 - \log[HCO_3^-], \tag{6.20}$$

and

$$p[Ca^{2+}] = 3.30 - \log[Ca^{2+}]. \tag{6.21}$$

K_{SP} and K_d are temperature dependent so that the temperature of the system under consideration must be specified. The temperature relationship of (pK_d–pK_{SP}) can be approximated by

$$pK_d - pK_{SP} = 2.586 - 2.621 \times 10^{-2}T, \tag{6.22}$$

where T is the water temperature in the irrigation system, in centrigrade.

The activity coefficients γ_{Ca} and γ_{HCO3} are also temperature dependent but in the 0–50 °C range the effect is negligible. However, the ion activity coefficient is concentration dependent through the ionic strength of the solution. According to Bower et al. (1965), the ionic strength (μ) of most irrigation water is related to its total dissolved salt concentration by

$$\mu = (1.3477C + 0.5355)/1000, \qquad (6.23)$$

where C is the concentration, mmol(charge)/l. Nakayama and Bucks (1985) calculated the following relationship

$$p(\gamma_{Ca}\gamma_{HCO_3}) = 7.790 \times 10^{-2} + 2.160 \times 10^{-2}C - 5.477 \times 10^{-4}C^2, \qquad (6.24)$$

where C is total dissolved salts in mmol(charge)/l.

The water in the various parts of the irrigation system may attain different temperatures during an irrigation day, e.g., in buried pipes the water temperature may be lower by more than 25 °C than that in a nonflowing black polyethylene lateral line used in drip irrigation systems (Nakayama and Bucks 1985). Thus LSI should be evaluated separately for the different components of the system. In steel, aluminum and asbestos-cement pipelines, moderate calcareous deposits formed under high water velocity can act as a protective coating against corrosion.

A spontaneous change in pH can occur in water containing large amounts of dissolved CO_2, brought about by a high hydraulic pressure. A decrease in this pressure once the water encounters the atmospheric pressure releases CO_2, causing a rise in pH and precipitation of dissolved $CaCO_3$.

Scale deposition problems are more likely to occur when water pH > 7.5 and bicarbonate > 5 mmol(−)/l. A simple way to qualitatively evaluate irrigation water for scale deposition is to fill a clean glass bottle with the water, raise its pH to approx. 9.5 by adding NH_4OH and let it stand for several hours. After slightly shaking the bottle, the precipitated solid calcium carbonate can be observed by directing a light beam onto the bottom of the bottle. The thickness of the deposit indicates the severity of the problem. Scale deposition is enhanced owing to pH increase when urea or anhydrous or liquid ammonia is applied through the irrigation system. For waters having pH > 7.5, it is recommended that sulfuric, hydrochloric or phosphoric acids be added to maintain the pH close to, but not lower than 6.5. The amount of acid addition to lower the pH is based on acid titration of the water.

Once scale is formed, the precipitate can be redissolved by reversing the conditions, but this occurs at a much slower rate than precipitation. Intermittent injection of acid requires lowering of the pH to approx 4.0, filling up the lines for 30–60 min, then letting the acidified water remain in the lines for 24 h, after which the lines must be flushed well. Injection of acid should always start with the system already in operation. Acid should not be added to asbestos-cement or cement lines. Calcium carbonate precipitation problems are more relevant to drip irrigation systems than to sprinkler systems, but may occur whenever small orifices are present.

6.6 Fertilizer and Chemical Injection Systems

Fertilization and application of chemicals through the irrigation system is possible with all sprinkler and drip irrigation methods and has become common practice for most agricultural crops. It is often called fertigation or chemigation.

Fertigation saves labor and costs. The operation of a fertilizer injection system is simple, convenient and fast. An injection system is, generally, also suitable for application of herbicides and pesticides, for soil-borne diseases and pests, and for acids and hypochlorites, which are important in the maintenance of the distribution systems, especially drip systems. By fertigation, timing of the fertilizer application is not dependent on soil surface moisture conditions, and soil compaction by tractor travel is thus avoided.

The application of fertilizer with the irrigation water results in a more suitable nutrient distribution within the root zone. This, in turn, leads to more efficient utilization. Another advantage is the ease of timing the quantity of nutrient application according to plant needs at different growth stages.

Fertigation requires fertilizers to be in liquid form or completely soluble in water. Some chemicals applied through the irrigation system are corrosive to metal parts of the system, e.g., galvanized iron is attacked by ammonium nitrate or sulphate and by phosphoric acid. Drip irrigation systems are prone to clogging by reaction of certain fertilizers or chemicals with water of different qualities. The seriousness of such problems may be evaluated by the following simple test. The fertilizer to be examined is mixed with a sample of the irrigation water, in a glass container, at a concentration similar to that which would be flowing through the distribution system. Then the glass container is held covered in a dark place for approx. 24 h. The container bottom is then examined with a light beam to determine whether any precipitate is formed.

Injection of chemical solutions into an irrigation pipeline, in which water is under pressure, requires the use of a device which can either use the energy of the flowing water or have an external energy source. There is a large variety of available injection equipment. The selection is based on the specific irrigation system, on the desired precision of application, reliability, convenience of operation and cost. The size or capacity of the injection system will depend on the concentrations, rate and frequency of application.

A most commonly used injection system is the Venturi pump. It operates on the principle of introducing a constriction into the water pipeline in which water flow is accelerated. The accelerated flow results in the creation of a pressure differential (vacuum) sufficient to suck the fertilizer solution from an open container into the irrigation pipelines, at a fixed dilution ratio. The Venturi pump system is generally installed in a bypass to reduce pressure head loss and to facilitate portability. These are simple systems with no moving parts and are relatively inexpensive. A wide assortment of injection pumps powered by external sources is also available. The most common are those of positive pumping, such as piston, gear, diaphragm or roller pumps, but suction pumps are also available. The power source can be an internal combustion engine, an electric motor or a hydraulic motor. The main advantage of these systems is the ease of integrating them into a central automatic computerized controller.

6.7 Filters

6.7.1 General Considerations

Suspended matter in the irrigation water is problematic in all irrigation systems. Drip and low-volume sprinkler distribution systems are particularly vulnerable to clogging, causing uneven water distribution. Suspended particles may also plug water passages and openings in the system's appurtenances, causing malfunction, e.g., automatic valves may not close as programmed and flow meters may give false readings. The abrasive action of sand, silt, small aquatic animals or weed seeds may damage diaphragms or impellers in pumps, flow meters, etc., causing reduced efficiency of the system (Adin 1986).

A list of contributors to the clogging of drip systems is given in Table 6.1 (Bucks et al. 1979). They are also troublesome in other irrigation systems, particularly in automatically controlled ones. Most of them are found in effluent, especially after storage. Often these clogging factors are interrelated and the problem is aggravated when more then one kind of contaminant is present in the water, which is usually the case. On the other hand, controlling one hazard may also alleviate problems caused by another. In general, two ways are available for improving irrigation water quality for reducing clogging and malfunction

Table 6.1 Principal physical, chemical and biological contributors to clogging of trickle systems (After Bucks et al. 1979)

Physical (Suspended solids)	Chemical (Precipitation)	Biological (Bacteria and algae)
Inorganic particles	Calcium or magnesium carbonate	Filaments
Sand		
Silt		Slimes
Clay	Calcium sulfate	
Plastic		Microbial decomposition:
	Heavy metal	
Organic particles:	Hydroxides, carbonates, silicates and sulfides	Iron
Aquatic plants (phytoplankton/algae)		Sulfur
		Manganese
Aquatic animals (zooplankton)	Oil or other lubricants	
Bacteria		
	Fertilizers:	
	Phosphate	
	Aqueous ammonia	
	Iron, copper, zinc, manganese	

hazards, namely, filtration and chemical treatment (chlorination and acidification). Applying only one method cannot always guarantee the efficient performance of an irrigation system, but a combination of the two can overcome most problems.

Installation of a filter is necessary in all effluent irrigation systems. Trash racks or coarse screens are usually installed before water pump intake. For sprinkler irrigation systems, the practice is to remove particles larger than one-third to one-half of the smallest orifice size, and in low pressure drip systems it is generally recommended to filter out particles as small as one-tenth to one-seventh of the emitter opening.

Filters used in irrigation systems may be classified into hydrocyclons; or centrifugal separators, and media and screen filters. Each type of filter is effective for a certain water quality as defined by the nature of the suspended solids, their size range, and their concentration. Success in selecting the right filtration system depends very much on how well the problem has been identified, and this depends on experience. Thus, it is frequently suggested, especially for drip systems, that two or more types of filters be installed in series. However, too much filtration can cause problems in the filtration system itself. Blockage of fine screens demands too much labor and water for backwashing and cleaning, or, in the case of automatic flushing filters, the cycle can become so short that failure of the irrigation system can occur. The aim of filtration of irrigation water should be to remove the larger-sized particles, allowing the final suspended load to be in the range that the delivery and distribution systems can tolerate for long-term operations. Adin and Elimelech (1989), concluded that for the evaluation of filter ability of waste water effluent, particle size distribution is more informative than TSS or turbidity.

A brief description of the various filter types is given below.

6.7.2 Hydrocyclon (Centrifugal) Filters

In these filters (Fig. 6.5), water enters a conical body (2) at the top (1) tangentially and creates a downward spinning of the water (5). The centrifugal forces thus produced throw solid particles heavier than water onto the wall which are carried in a spiral motion to the bottom (6) of the collecting tank (4), which is connected to the main body (3). The clean water rises up in the centered vortex (7) to the outlet and into the irrigation system. The collecting tank can be drained periodically or continuously through the drainage valve (8). Hydrocyclons are simple devices and have no moving or replaceable parts, and once installed in line, can operate unattended. The pressure head loss between inlet and outlet, which is the energy required to activate the vortex, is constant and may be as low as 0.5 m or as high as 6 m, depending on flow. They can effectively separate sand particle larger than 75 μm, or with special design even 50 μm. As a rule, separation improves as the pressure drop increases, and as the hydrocyclon diameter decreases or as flow rate increases. Hydrocyclons can be designed for any water flow rate and sediment load and for any inlet pressure. However, to achieve

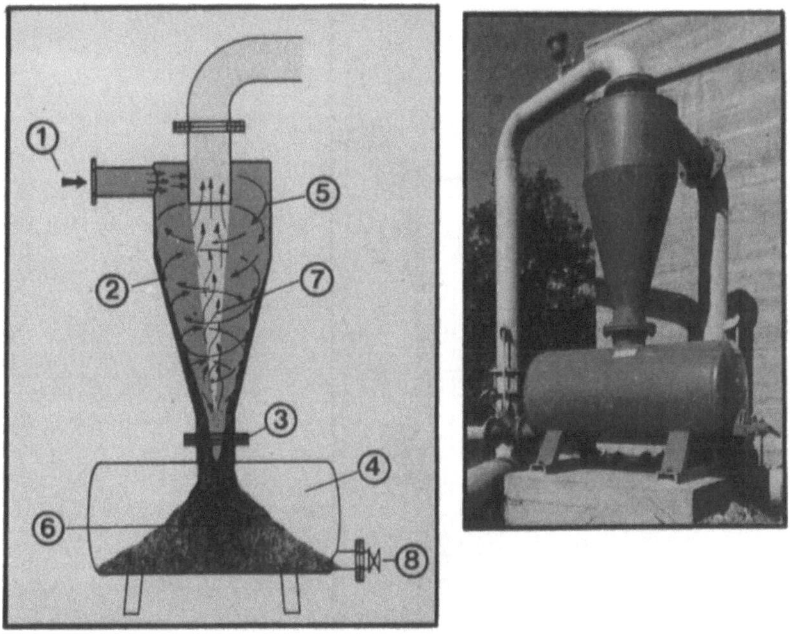

Fig. 6.5 Hydrocyclon filter: *1* inlet; *2* conical body; *3* connector; *4* collecting tank; *5* water spin; *6* particles; *7* vortex; *8* drainage valve (Netafim Company, Israel)

maximum performance they must operate within their prescribed flow range. When the flow rate varies beyond acceptable limits, it is preferable to install smaller units in parallel with individual or common collecting tanks. Such an arrangement is also possible for increasing the capacity when using small units, which can remove smaller particles.

6.7.3 Screen Filters

Filters of this type (Fig. 6.6) work as strainers. They can be used effectively to remove particles of at least 75 μm from irrigation water with a moderate level of suspended solids. But most frequently they are used in irrigation systems requiring moderate filtration and as a backup for media filters in drip irrigation systems.

Cylindrical screens made of plastic or stainless steel are enclosed in a sturdy plastic or corrosive protected metal body. The mode of water entry, circulation and exit varies according to the model.

The screens are characterized by the equivalent diameter of the opening and the effective area ratio, which is the ratio of the total area of the perforations to the total screen area. This ratio is generally in the range of 0.25–0.40, typically 0.35. Another commonly used criterion is the mesh number, which is the number of

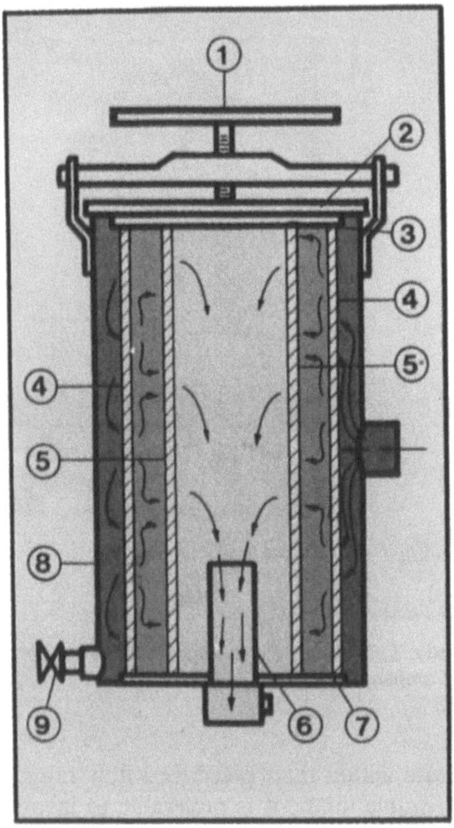

Fig. 6.6 Screen filter: *1* filter shutter; *2* cover; *3* upper seal; *4* exterior filter element; *5* interior filter element; *6* PVC nipple; *7* lower seal; *8* filter body; *9* drainage valve (Netafim Company, Israel)

openings per linear inch. For a screen with an effective area ratio of 0.35, the equivalent diameter (in mm) is approximately 15 divided by the mesh number, and the wire thickness (in mm) is 10 divided by the mesh number. The most common mesh in sprinkler irrigation systems is 20–60 (850–250 μm), and in drip irrigation 100–140 (150–106 μm), although filters with 200 mesh (75 μm) screens are frequently recommended.

An important parameter characterizing screen filters is the filter ratio, which is the ratio between the total screen area to the water inlet area. The filter ratio should be larger than 4 for the lowest flow rate.

A clean screen has virtually no pressure loss, but, as the solids accumulate on the screen, the pressure head loss increases. Since the pressure loss is exponentially related to the volume of accumulated solids, the pressure differential increases very gradually at first, then, with prolonged operation, more rapidly, until it causes reduced flow rates and ultimately screen collapse, tear or failure. It is generally recommended that the filter screen is cleaned when the differential pressure head reaches approximately 3 m. The filter should be equipped with pressure gauges at the inlet and outlet ends to indicate the screen condition (when the screen tears the differential pressure is approx. 0) and timing of flushing. Plastic screens should be supported to prevent deformations by high differential pressure.

In small, simple screens filters, the screen must be removed and washed by hand. In others it can be done by backwashing or flushing with water at very high flow rate, either manually or automatically. The irrigation system must be designed to assure that adequate pressure is available to achieve the minimum flushing flow rate recommended by the manufacturer. Nevertheless, periodic shutdown, disassembly and thorough cleaning, preferably using a brush, is recommended.

Algae, which grow profusely in effluent reservoirs, may cause serious problems with screening filtration, due to their tendency to become entangled within the screen mesh.

In recent years, filters with grooved discs or rings, as an alternative to conventional screens, have become available for irrigation systems. The disc plates are pressed together during operation and the grooves control the opening size. The long opening path is intended to provide a larger volume for the accumulated strained particles and improve the filter performance. During flush mode or cleaning, the discs are separated to facilitate removal of the accumulated solids (Fig. 6.7).

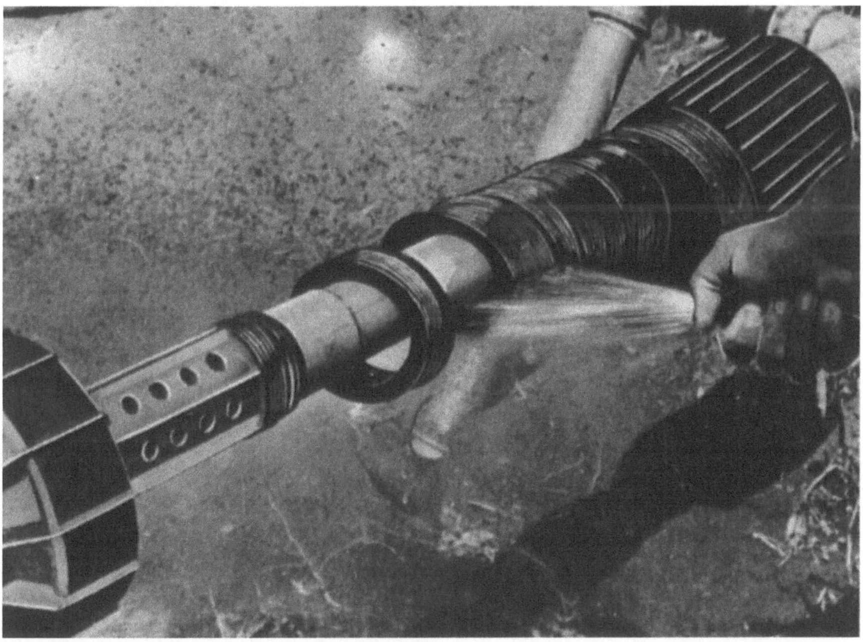

Fig. 6.7 Disc filter during cleaning (Arkel, Israel)

6.7.4 Gravel or Sand Media Filters

These filters in which water percolate through a packed bed of fine gravel or sand are the most versatile filtration alternative. They can effectively retain both organic and inorganic suspended particles of at least a few of microns in size. Gravity operating sand filters are widely used in surface water supply systems and in final treatment stages of sewage effluent. Pressurized, high-flow gravel filters are very popular in drip irrigation systems (Fig. 6.8). Incoming flow rates should not exceed $0.1\,l\,min^{-1}\,cm^{-2}$ of the media surface area, to prevent channeling. Almost the entire depth of the packed gravel (approx. 30 cm) is actively used in retaining the suspended solids. The pressure head loss of a clean gravel filter is low (1–1.2 m). Media filters are cleaned when the differential pressure head loss reaches 6–8 m by reversing the flow so that the bed is lifted and expanded, allowing the trapped sediments to be released through turbulent mixing. Thus, filters should be designed with approximately 10 cm extra height and for an increased flow rate of approximately 65%. Since finer bed particles tend to accumulate on the top, it is advisable to scrape off the top layer after the first few backwashings. (This also helps to get rid of algae clogging the sand surface.)

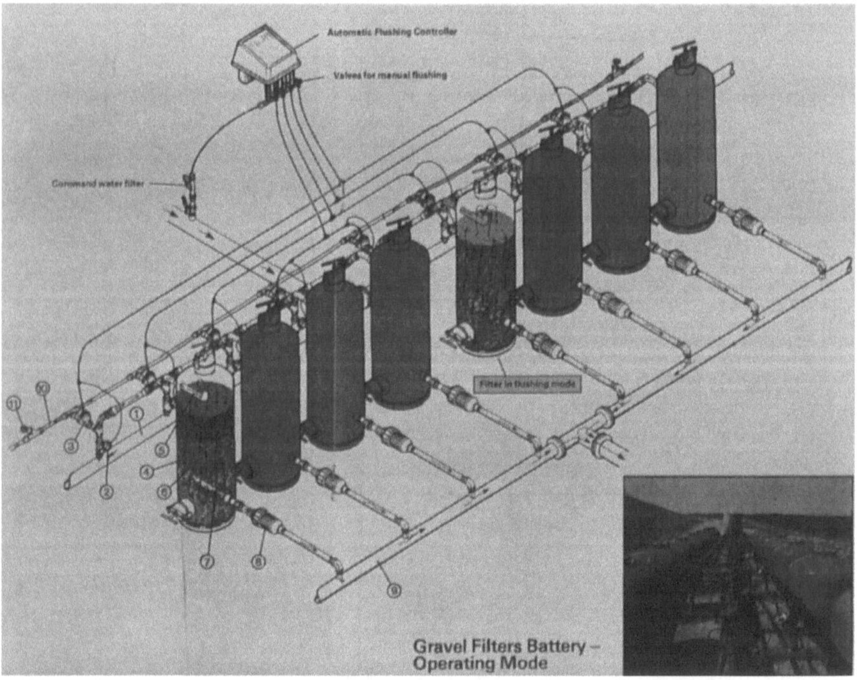

Fig. 6.8 Gravel filter: *1* inlet manifold-nonfiltered water; *2* hydraulic inlet valve; *3* hydraulic drainage valve; *4* gravel filter; *5* filter inlet; *6* gravel substrate; *7* filter element; *8* screen filter; *9* outlet manifold; *10* drainage pipe; *11* drainage valve (Netafim Company, Israel)

Backwashing can be accomplished manually, but automatic systems based upon timers (generally over 90 min), and/or differential pressure are more common. Dual filter systems are usually used, so that the filtered water from one is passed through the other for backwashing to prevent unfiltered water from entering the drip system. Backwashing takes 1–2 min. A screen filter is usually installed as a backup after the media filter. Media filters provide favorable conditions for bacterial growth, especially with effluent, causing aggregation of the media particles. Cementing of the media particles with clay and silt and by calcium carbonate precipitation can also occur. Thus, it is recommended to occasionally mix the bed and add chlorine compounds. Chlorination for 1/2 h with 10–20 mg/l free residual chlorine before the irrigation system is shut off at the end of the irrigation season, is also advisable.

Filtration media, gravel and sand particles should not consist of widely variable particle sizes, i.e., not more than + 20% of the mean size. The particle size usually recommended for drip irrigation range from 0.47–1.40 mm.

6.8 Irrigation System Maintenance

Complete and timely maintenance is essential for successful performance of an effluent irrigation system. Proper maintenance will extend the life of the system, reduce operating costs and minimize unplanned shutdowns.

Flow rate measurements are extremely useful in assessing the need for irrigation system maintenance as well as monitoring its performance. They facilitate detection of pipeline leaks and breaks, inadequate pressure regulation, improper adjustment of valves, clogging and blockage in various parts of the system. Knowledge of water flow rate and volume is also important for good irrigation management.

Several types of flow meters are available for measuring the flow volume or volumetric flow rate.

The most common meters for measuring flow in pipelines are the *mechanical* or *propeller meters*. They consist basically of a multibladed propeller which revolves at a speed roughly proportional to the water flow rate, and a mechanical system for recording and displaying the flow rate and volume. Generally, the flow volume is read on a digital accumulator and the flow rate is read from a needle on a dial. These meters are widely used because each consists of a separate complete unit and they present the least problems in installation. The effective range of low-to-high flow rate for an individual meter is about 1–10. The head loss through most mechanical meters is moderated and their accuracy is 5–2%, depending on the type of propeller. They are commercially available for a wide range of pipe sizes and operate within a wide range of flow velocities.

Flow rates can be obtained from measurements of the pressure differential created when the flowing water is directed through a constriction. The most common constriction is an orifice (for pipes, Venturi meters are also common). The pressure difference for a given orifice is proportional to the square of the water flow rate. Manometers, bourdon gauges and transducers are usually used to measure the pressure difference. They may be calibrated to read flow rates directly.

Orifice meters are available for measuring flow rates in pipes and in open canals. In general, the measurement of open canal flow rate is more troublesome and less accurate than measurement of flow rate in a fully flowing pipe.

Particularly suited for measuring flow in irrigation canals are flumes, the most common of which are the Parshall flumes, the H-flumes and the cut-throat flume. Sharp-crested weirs are also widely used in irrigation canal systems.

Other flow rate meters, such as Venturi meters, electronic and sonic flow meters, are available for pipeline systems. These are more accurate but are generally expensive, and therefore are mainly used on main supplying pipelines.

Pipelines in the irrigation system should be inspected periodically for leaks while the system is in operation. If the irrigation system is located in freezing climate, drain valves should be installed to drain the pipe system. At the end of the irrigation season, the pipeline system should be flushed (if possible with fresh water or chlorinated effluent) and drained, and all valves should be opened. Effluent left in the pipe may become septic and/or suspended matter may settle and harden at low points in the line and cause clogging problems. It is also advisable to inspect pipes for corrosion and apply protective measures when necessary.

Open canals should also be inspected periodically for damages. They need cleaning to remove sediment deposits. Unlined canals should be inspected for erosion damage, for rodents that can cause piping and washouts and for excess vegetation. A fair amount of grass growing along the canal banks is desirable for bank stabilization. Metal structures in the system should be painted or applied with a protective coating. Metal gates should be lubricated and, where possible, left partially open during winter.

Pumps, motors, engines and injection equipment should be checked during the season for leaks and excessive noise, temperature, vibrations and pressures. Wherever lubrication is required, manufacturer recommendations should be closely followed. Most system equipment come with instruction manuals which usually contain recommendations for off-season maintenance and a trouble-shooting section, and should be consulted.

References

Adin A (1986) Problems associated with particulate matter in water reuse for agricultural irrigation and their prevention. Water Sci Tech 18:185–195

Adin A, Elimelech M (1989) Particle filtration for wastewater irrigation. J Irrig Drain Div ASCE 115:474–483

Adriano DC, Pratt PF, Takatori FH (1972) Nitrate in unsaturated zone of alluvial soil in relation to fertilizer nitrogen rate and irrigation level. J Environ Qual 1:418–422

Alexander M (1977) Introduction to soil microbiology. John Wiley & Sons, New York

Alexander M (1981) Biodegradation of chemicals of environmental concern. Science 211:132–138

Ali I (1987) Wastewater criteria for irrigation in arid regions. J Irrig Drain Div ASCE 113:173–183

Allaway WH (1977) Food chain aspects of the use of organic residues. In: Elliot LF, Stevenson FJ (eds) Soils for management of organic wastes and waste waters. Am Soc Agron, Madison, pp 283–298

Amoozegar-Fard A, Nielsen DR, Warrick AW (1982) Soil solute concentration distributions for spatially varying pore water velocities and apparent diffusion coefficients. Soil Sci Soc Am J 46:3–9

Arar A (1987) The role of the use of sewage effluent for irrigation development in the Near East Region. In: Wunderlich WO, Prins JE (eds) Water for the future, water resources developments in perspective. Balkema, Rotterdam Boston, pp 245–260

Asano T, Pettygrove GS (1987) Using reclaimed municipal wastewater for irrigation. Cal Agric 41(3–4):15–18

Asano T, Smith RG, Tchobanoglous G (1985) Municipal wastewater: treatment and reclaimed water characteristics. In: Pettygrove GS, Asano T (eds) Irrigation with reclaimed municipal wastewater – a guidance manual. Lewis, Chelsea, pp 2.1–2.26

Avnimelech Y (1984) Behavior of phosphate in the unsaturated zone. In: Yaron B, Dagan G, Goldshmid J (eds) Pollutants in porous media. Springer, Berlin Heidelberg New York, pp 68–78

AWWA (American Water Workers Association) (1981) Conference/tour report. AWWA Res Found, 6666 Quincy Ave, Denver, Col. In: Municipal Wastewater Reuse News 45:3–11

Ayers RS, Westcot DW (1985) Water quality for agriculture, Rev edn. FAO Irrig Drain Pap 29 Rome 174 pp

Baier DC, Fryer WB (1973) Undesirable plant responses with sewage irrigation. J Irrig Drain Div ASCE 99:133–141

Bailey LD, Soper RJ (1985) Potassium nitrition of rape, flax, sunflower, and safflower. In: Munson RD (ed) Potassium in agriculture. Am Soc Agron, Madison, pp 765–798

Barrow NJ (1980) Differences among some North American soils in the rate of reaction with phosphate. J Environ Qual 9:644–648

Bar-Yosef B (1977) Trickle irrigation and fertilization of tomatoes on sand dunes: water, N and P distribution in the soil and uptake by plants. Agron J 69:486–491

Bar-Yosef B, Kafkafi U (1978) Phosphate desorption from kaolinite suspensions. Soil Sci Soc Am J 42:570–574

Bassett DM, McKenzie AJ (1978) Plant analysis as a guide to cotton fertilization. In: Reisenauer HM (ed) Soil and plant testing in California. Div Agric Sci Univ Cal Bull 1879

Bausum HT, Schaub SA, Bates RE, McKim HL, Schumacher PW, Brockett BE (1983) Microbiological aerosols from a field-source wastewater irrigation system. J Water Pollut Control Fed 55:65–75

Beaton JD, Sekhon GS (1985) Potassium nutrition of wheat and other small grains. In: Munson RD (ed) Potassium in agriculture. Am Soc Agron Madison, pp 701–752

Benami A, Ofen A (1983) Irrigation engineering: sprinkler, trickle, surface irrigation-principles, design and agricultural practices. IESP Haifa, Israel/IIIC, POB 49, 50 250 Bet Dagan, Isr, 257 pp

Bernstein L (1961) Osmotic adjustment of plants to saline media in steady state. Am J Bot 48:904–918

Bernstein L (1964) Effects of salinity on mineral composition and growth of plants. Plant Anal Fertil Probl 4:25–45

Bernstein L (1974) Crop growth and salinity. In: van Schilfgaarde J (ed) Drainage for agriculture. Am Soc Agron Madison. Agronomy 17:39–54.

Bernstein L, Francois LE (1973) Leaching requirement studies: sensitivity of alfalfa to salinity of irrigation and drainage waters. Soil Sci Soc Am Proc 37:931–943

Bertsch PM, Thomas GW (1985) Potassium status of temperate regions soils. In: Munson RD (ed) Potassium in agriculture. Am Soc Agron Madison, pp 131–162

Bielorai H, Feigin A, Hadas A (1980) Response of cotton to irrigation with treated municipal effluents. In: Banin A, Kafkafi U (eds) Agrochemicals in soils. Pergamon, Oxford, pp 395–403

Bielorai H, Vaisman I, Feigin A (1984) Drip irrigation of cotton with treated municipal effluents. 1: Yield response. J Environ Qual 13:231–234

Bingham FT, Page AL, Mahler RJ, Ganje TJ (1975) Growth and cadmium accumulation of plants grown on a soil treated with a cadmium-enriched sewage sludge. J Environ Qual 4:207–211

Bingham FT, Page AL, Mahler RJ, Ganje, TJ (1976) Yield and cadmium accumulation of forage species in relation to cadmium content of sludge-amended soil. J Environ Qual 5:57–60

Black CA (1968) Soil-plant relationships, 2nd edn. John Wiley & Sons, New York

Blaney, HF, Criddle WD (1950) A method of estimating water requirement in irrigated areas from climatological data. USDA Soil Conserv Serv Tech Publ 96

Blum D, Feachem RG (1985) Health aspects of night soil and sludge use in agriculture and aquaculture. An epidemiological perspective. International Reference for Water Disposal, Dubendorf, Switzerland

Boaz M, Hausenberg I, Pozin Y (1976) Salinity survey in Israel. In: Dregne HE (ed) Managing saline water for irrigation. Proc Int Salinity Conf, Texas Tech College, Lubbock, pp 388–399

Bohn HL, McNeal BL, O'Connor GA (1985) Soil chemistry, 2nd edn. John Wiley & Sons, New York

Bole JB, Bell RG (1978) Land application of municipal sewage waste water. Yield and chemical composition of forage crops. J Environ Qual 7:222–226

Bolt GH (1978) Transport and accumulation of soluble soil components: In: Bolt GH, Bruggenwert MGM (eds) Soil chemistry. A: Basic elements. Elsevier, Amsterdam, pp 126–140

Bond RG, Straub CP (eds) (1974) Wastewater treatment and disposal control, vol 4. CRC Handbook of environmental control. CRC, Boca Raton 905 pp

Booker LJ (1974) Surface irrigation. FAO Agric Dev Pap 95, Rome

Bouwer H (1970) Groundwater recharge design for renovating wastewater. J Sanit Eng Div Am Civ Eng 96:59–74

Bouwer H (1974) Developing drainage design criteria. In: van Schilfgaarde J (ed) Drainage for agriculture. Am Soc Agron Madison. Agronomy 17:67–79

Bouwer H (1982) Wastewater reuse in arid areas. In: Middelbrooks EJ (ed) Water reuse. Science Publisher, Ann Arbor, pp 137–180

Bouwer H, Chaney RL (1974) Land treatment of wastewater. Adv Agron 26:133–176

Bouwer H, Idelovitch E (1987) Quality requirements for irrigation with sewage water. J Irrig Drain Div ASCE 113:516–535

Bouwer H, van Schilfgaarde J (1963) Simplified method of predicting fall of water table in drained land. Trans Am Soc Agric Eng 6:288–291, 296

Bouwer H, Lance JC, Riggs MS (1974a) High rate land treatment II. Water quality and economic aspects of the Flushing Meadows Project. J Water Pollut Control Fed 46:844–859

Bouwer H, Rice RC, Escarcega ED (1974b) Infiltration and hydraulic aspects of the Flushing Meadows Project. J Water Pollut Control Fed 46:835–843

Bower CA, Wilcox LV, Akin GW, Keyes MG (1965) An index of the tendency of $CaCO_3$ to precipitate from irrigation water. Soil Sci Soc Am Proc 29:91–92

Bower CA, Ogata G, Tucker JM (1969) Root zone salt profiles and alfalfa growth as influenced by irrigation water salinity and leaching fraction. Agron J 61:783–785

Bower CA, Ogata G, Tucker JM (1970) Growth of Sudan and tall fescue grasses as influenced by irrigation water and leaching fraction. Agron J 62:793–794

Bremner JM, Shaw K (1958) Denitrification in soil. 2. Factors affecting denitrification. J Agric Sci 51:39–52

Bresler E (1977) Trickle-drip irrigation: principles and application to soil-water management. Adv Agron 29:343–393

Bresler E, Dagan G (1979) Solute dispersion in unsaturated heterogeneous soil at field scale. 2: Applications. Soil Sci Soc Am J 43:467–472

Bresler E, Dagan G (1981) Convective and pore scale dispersive solute transport in unsaturated heterogeneous fields. Water Resour Res 17:1683–1693

Bresler E, McNeal BL, Carter DL (1982) Saline and sodic soils: principles-dynamics-modeling. Springer, Berlin Heidelberg New York

Broadbent FE, Reisenauer HM (1985) Fate of wastewater constituents in soil and groundwater: nitrogen and phosphorus. In: Pettygrove GS, Asano T (eds) Irrigation with reclaimed municipal wastewater – a guidance manual. Lewis, Chelsea, pp 12.1–12.16

Bryan FL (1977) Diseases transmitted by foods contaminated by wastewater. J Food Protect 40:45–56

Bucks DA, Nakayama FS, Gilbert, RG (1979) Trickle irrigation water quality and prevention maintenance. Agric Water Manag 2:149–162

Bucks DA, Nakayama FS, Warrick AW (1982) Principles, practices and potentialities of trickle (drip) irrigation. Adv Irrig I: 219–298

Burau RG, Sheikh B, Cort RP, Cooper RC, Ririe D (1987) Reclaimed water for irrigation of vegetables eaten raw. Cal Agric 41(7–8):4–7

Burford JR, Bremner JM (1975) Relationship between the denitrification capacities of soils and total, water soluble and readily decomposable soil organic matter. Soil Biol Biochem 7:389–394

Burge WD, Marsh PB (1978) Infectious disease hazards of landspreading sewage wastes. J Environ Qual 7:1–9

Burman RD, Nixon PR, Wright JL, Pruitt WO (1980) Water requirement. In: Jensen ME (ed) Design and operation of farm irrigation systems. ASAE Monograph No 3. Am Soc Agric Eng, St Joseph, Mich, pp 189–232

Camann DE, Moore BE, Harding HJ, Sorber CA (1988) Microorganism levels in air near spray irrigation of municipal wastewater: the Lubbock infection surveillance study. J Water Pollut Control Fed 60:1960–1970

Carter JN, Westermann DT, Jensen ME (1976) Sugarbeet yield and quality as affected by nitrogen level. Agron J 68:49–55

CAST (Council for Agricultural Science and Technology – ed) (1976) Application of sewage sludge to cropland: appraisal of potential hazards of the heavy metals to plants and animals. Rep 64, Ames Iowa 63 pp

CAST (Council for Agricultural Science and Technology – ed) (1985) Agriculture and ground water quality. Rep 103, Ames, Iowa 59 pp

Castro CL, Rolston DE (1977) Organic phosphate transport and hydrolysis in soil: theoretical and experimental evaluation. Soil Sci Soc Am J 41:1085–1092

Chaney RL (1989) Toxic element accumulation in soils and crops: protecting soil fertility and agricultural food-chains. In: Bar-Yosef B, Barrow NJ, Goldshmid J (eds) Inorganic contamination of the vadose zone. Springer Ecol Stud 74:140–158

Chaney RL, Giordano PM (1977) Microelements as related to plant deficiencies and toxicities. In: Eliott LF, Stevenson FJ (eds) Soils for management of organic wastes and waste waters. Am Soc Agron, Madison, pp 235–279

Chaney RL, Bruins RJF, Baker DE, Korchak RF, Smith JE Jr, Cole DW (1987) Transfer of sludge-applied trace elements to the food chain. In: Page AL, Logan TJ, Ryan JA (eds) Land application of sludge, food chain implications. Lewis, Chelsea, pp 67–99

Chang AC, Page AL (1983) Fate of trace metals during land treatment of municipal wastewater. In: Page AL, Gleason TL, Smith JE, Iskander IK, Sommers LE (eds) Utilization of municipal wastewater and sludge on land. Univ Press Cal, Riverside, pp 107–122

Chang AC, Page AL (1985) Fate of wastewater constituents in soil and groundwater: trace organics. In: Pettygrove GS, Asano T (eds) Irrigation with reclaimed municipal wastewater – a guidance manual. Lewis, Chelsea, pp 15.1–15.20

Chang AC, Page AL, Bingham FT (1981) Reutilization of municipal wastewater sludges-metals and nitrate. J Water Pollut Control Fed 53:237–245
Chang AC, Page AL, Warneke JE, Johanson JB (1982) Effects of sludge application on the Cd, Pb and Zn levels of selected vegetable plants. Hilgardia 50:1–14
Chang Y-Hu (1968) Climate and agriculture. Aldine, Chicago
Chick H (1908) An investigation of disinfection. J Hyg 8:698
Clapp CE, Linden DR, Larson WE, Marten GC, Nylund JR (1977) Nitrogen removal from municipal wastewater effluent by a crop irrigation system. In: Loehr RC (ed) Land as a wastewater management alternative. Science Publisher, Ann Arbor, pp 139–150
Committee on Nitrate Accumulation in Soils (ed) (1972) Accumulation of nitrate. Nat Acad Sci, Washington, DC
Cooke GW (1972) Fertilizing for maximum yield. Lockwood, London
Cooper WJ (ed) (1981) Chemistry in water reuse, vol 1. Science Publisher, Ann Arbor; and Butterworth, London
Cowan JP, Johnson PR (1984) Reuse of effluent for agriculture in the Middle East. In: Reuse of sewage effluent. Telford, London, pp 107–127
Crites RW (1974) Irrigation with wastewater at Bakersfield, California. In: Wastewater use in the production of food and fiber. Proc US Environ Protect Ag, EPA-660/2-74-041, pp 229–239
Crites RW (1985) Site characteristics. In: Pettygrove GS, Asanop T (eds) Irrigation with reclaimed municipal wastewater – a guidance manual. Lewis, Chelsea, pp 4.1–4.19
Crook J (1985) Health and regulatory considerations. In: Pettygrove GS, Asano T (eds) Irrigation with reclaimed municipal wastewater – a guidance manual. Lewis, Chelsea, pp 10.1–10.49
Dag J, Feigin A, Giskin M, Golan S, Jerome G, Davidov S, Keinan M (1984) Response of cotton to effluent irrigation in a grumusol soil in the Yizre'el Valley; 1976–1982 experiments. 1: Cotton yield. Agric Res Org Bet Dagan, Isr. Pamphl 228:47 pp (Hebrew with English summary)
Dagan G (1986) Statistical theory of groundwater flow and transport: pore to laboratory, laboratory to formation and formation to regional scale. Water Res 17:1683–1693
Dagan G, Bresler E (1979) Solute dispersion in unsaturated heterogeneous soil at field scale. 1: Theory. Soil Sci Soc Am J 43:461–467
Dagan G, Bresler E (1983) Unsaturated flow in spatially variable fields. 1: Derivation of models of infiltration and redistribution. Water Resour Res 19:413–420
Dalal RC (1977) Soil organic phosphorus. Adv Agron 29:83–117
Dasberg S, Bresler E (1985) Drip irrigation manual. IIIC. Volcani, Bet Dagan, Isr 50250, 108 pp
Dasberg S, Dalton FN (1985) Time domain reflectometry field measurements of soil water content and electrical conductivity. Soil Sci Soc Am J 49:293–297
Davidson JM, Rao PSC Nkedi-Kizza P (1983) Physical processes influencing water and solute transport in soil. In: Nielsen DW, Elrick DW, Tanji KK (eds) Chemical mobility and reactivity in soils systems. Soil Sci Soc Am Spec Publ 11:35–47
Day AD (1973) Recycling urban effluent on land using annual crops. In: Recycling municipal sludges and effluents on land. In: Proc Joint Conf Natl Assoc State Univ Land-Grant Colleges, Washington, DC, pp 155–160
Day AD, Kirkpatrick RM (1973) Effects of treated municipal wastewater on oat forage and grain. J Environ Qual 2:282–284
Day AD, Tucker TC (1959) Production of small grains pasture forage using sewage effluent as a source of irrigation water and plant nutrients. Agron J 51:569–572
Day AD, Tucker TC (1960) Hay production of small grains utilizing city sewage effluent. Agron J 52:238–239
Day AD, Tucker TC (1977) Effects of treated municipal wastewater on growth, fiber, protein and amino acid content of sorghum grain. J Environ Qual 6:325–327
Day AD, Stroehlein JL, Tucker TC (1972) Effects of treatment plant effluent on soil properties. J Water Pollut Control Fed 44:372–375
Day AD, McFadyen JA, Tucker TC, Cluff CB (1981) Effect of municipal wastewater on the yield and quality of cotton. J Environ Qual 10:47–49
Day AD, Swingle RS, Tucker TC, Cluff CB (1982) Alfalfa hay growth with municipal waste water and pump water. J Environ Qual 11:23–24

De Datta SK, Mikkelsen DS (1985) Potassium nutrition in rice. In: Munson RD (ed) Potassium in agriculture. Am Soc Agron Madison, pp 665–699

Dedrick AR, Erie LJ, Clemmens AJ (1982) Level basin irrigation. Adv Irrig 1:105–145

Doorenbos J, Kassam AH (1979) Yield response to water. FAO Irrig Drain Pap 33:193 pp

Doorenbos J, Pruit WO (1977) Crop water requirements. FAO Irrig Drain Pap 24:156 pp

Dowdy RH, Larson WE (1975) The availability of sludge-borne metals to various vegetable crops. J Environ Qual 4:278–282

Dowdy RH, Volk VV (1983) Movement of heavy metals in soils. In: Nelson DW, Elrick EE, Tanji KK (eds) Chemical mobility and reactivity in soil systems. Soil Sci Soc Am Spec Pub 11:229–240

Dowdy RH, Larson WE, Titrud JM, Latterell JJ (1978) Growth and metal uptake of snap beans grown on sewage sludge-amended soil: A four-year study. J Environ Qual 7:252–257

Eitan G (1989) Water reclamation for irrigation and open reservoirs as part of treatment. J Assoc Eng Archit Isr: Sept 1989:19 (in Hebrew with English Summary)

Eliott LF, Stevenson FJ (eds) (1977) Soils for management of organic wastes and waste waters. Soil Sci Soc Am, Am Soc Agron, Crop Sci Soc Am, Madison, 650 pp

Ellis BG (1974) The soil as a chemical filter. In: Sopper WE, Kardos LT (eds) Conference on recycling treated municipal wastewater through forest and cropland. US Environmental Protection Agency, EPA-660/2-740-003, pp 47–72

Ellis BG (1976) Analyses and their interpretation for wastewater application on agricultural land. In: Knezek BD, Miller RH (eds) Application of sludges and wastewaters on agricultural land: a planning and educational guide. Ohio Agric Res Dev Center, Wooster, Res Bull 1090:6.1–6.6

Ellis BG, Eriksson AE, Knezek BD, Wolcott AR, Bahr TG (1972) Impact of wastewater on soils. Inst Water Resour Mich State Univ, East Lansing, Tech Rep 30

Enfield CG (1981) Field evaluation of a phosphorus model. In: Iskander IK (ed) Modeling wastewater renovation, land treatment. John Wiley & Sons New York, pp 668–685

Enfield CG, Ellis R Jr (1983) The movement of phosphorus in soil. In: Nelson DW, Elrick EE, Tanji KK (eds) Chemical mobility and reactivity in soil systems. Soil Sci Soc Am Madison Spec Publ 11:93–107

Engelberg Report (1985) Health aspects of wastewater and excreta use in agriculture. In: rep rev meet environmental specialists and epidemiologists, Engleberg, Switz, July 1–14 1985. International Reference Center for Wastes Disposal, Dubendorf, Switzerland

Engelbrecht RS (1980) Comparative inactivation of viruses by chlorine application. Appl Environ Microbiol 40:249–256

EPA (US Environmental Protection Agency – ed) (1979) Criteria for classification of solid waste disposal facilities and practices. EPA Fed Register 44:53438–53464

EPA (US Environmental Protection Agency – ed) (1981) Process design manual for land treatment of municipal wastewater. US Environ Protect, Washington, DC, EPA 625/1-77-008

Evans KJ, Mitchell IG, Salau B (1979) Heavy metal contamination in soils irrigated by sewage and effect on the plant-animal system. Prog Water Technol 11:339–352

Fair GM, Geyer JC, Okun DA (1971) Water and wastewater engineering. John Wiley & Sons, New York

FAO (Food and Agricultural Organization – ed) (1985) Report on the proceedings of the FAO regional seminar on the treatment and use of sewage effluent for irrigation. In: Reg Commiss Land and water use in the Near East. Cyprus, 10–12 Oct 1985, FAO, Rome, 10 pp

Feagley SE, Hossner LR (1978) Ammonia volatilization reactions mechanism between ammonium sulfate and carbonate systems. Soil Sci Soc Am J 42:364–367

Feigenbaum S, Edelstein R, Shainberg I (1981) Release rate of potassium and structural cations from micas to ion exchange in dilute solutions. Soil Sci Soc Am Proc 45:501–506

Feigin A, Kipnis T (1980) Improving N uptake by Rhodes grass from treated municipal effluents, to reduce fertilizer requirements and prevent pollution. In: Banin A, Kafkafi U (eds) Agrochemicals in soils. Pergamon, Oxford, pp 387–394

Feigin A, Sagiv B (1975) The response of tomatoes for processing to nitrogen and phosphorus fertilization in a light brown loessial soil. Isr Agric Res Org, Div Sci Publ Prelim Rep 750

Feigin A, Shakib B (1972) Soil and plant analysis for forecasting needs for nitrogen, phosphorus and potassium fertilizers of tomatoes for processing. Div Sci Publ Prelim Rep 684. ARO, Volcani, Bet Dagan, Isr (in Hebrew)

Feigin A, Bielorai H, Dag Y, Kipnis T, Giskin M (1978) The nitrogen factor in the management of effluent-irrigated soils. Soil Sci 125:248–254

Feigin A, Bielorai H, Shalhevet J, Kipnis T, Dag J (1979) The effectiveness of some crops in removing minerals from soils irrigated with sewage effluent. Prog Water Technol 11:151–162

Feigin A, Feigenbaum S, Limoni H (1981) Utilization efficiency of nitrogen from sewage effluent and fertilizer applied to corn plants growing in a clay soil. J Environ Qual 10:284–287

Feigin A, Vaisman I, Bielorai H (1984) Drip irrigation of cotton with treated municipal effluents. 2: Nutrient availability in soil. J Environ Qual 13:234–238

Fenn LB, Kissel DE (1973) Ammonia volatilization from surface applications of ammonium compounds on clacareous soils. 1: General theory. Soil Sci Soc Am Proc 37:855–859

Finkel HJ (ed) (1982) Handbook of irrigation technology, vols 1 and 2. CRC, Boca Raton

Firestone MK (1982) Biological denitrification. In: Stevenson FJ (ed) Nitrogen in agricultural soils. Am Soc Agron, Madison. Agronomy 22:289–326

Fleisher Z, Kenig A, Ravina I, Hagin J (1987) Model of ammonia volatilization from calcareous soils. Plant Soil 103:205–212

Focht DD, Verstraete W (1977) Biochemical ecology of nitrification and denitrification. Adv Microbiol Ecol 1: 135–214

Frankenberger WT Jr (1985) Fate of wastewater constituents in soil and groundwater: pathogens. In: Pettygrove GS, Asano T (eds) Irrigation with reclaimed municipal wastewater – a guidance manual. Lewis, Chelsea, pp 14.1–14.25

Frenkel H, Vinten A, Mantell A (1984) Crop yield and water use under irrigation with saline water. First annual report to BARD. Volcani, Bet Dagan, Isr, 22 pp

Fuller WH, Tucker TC (1977) Land utilization and disposal of organic wastes in arid regions. In: Eliott LF, Stevenson FJ (eds) Soils for management of organic wastes and waste waters. ASA, SSSA, CSA, Madison, pp 473–489

Fuller WH, Warrick AW (1985) Soils in waste treatment and utilization, vols 1 and 2 CRC, Boca Raton

Gammon W (1957) The behavior of phosphorus and potassium fertilizer in Florida soils. Proc Soil Crop Sci Soc Fla 17:156–160

George MR, Pettygrove GS, Davis WB (1985) Crop selection and management. In: Pettygrove GS, Asano T (eds) Irrigation with reclaimed municipal wastewater – a guidance manual. Lewis, Chelsea, pp 6.1–6.18

Gerba CP (1983) Pathogens. In: Page HL, Gleason TL, Smith JE, Iskander IK, Sommers LE (eds) Utilization of municipal wastewater and sludge on land. Univ Press Cal, Riverside, pp 147–195

Gerba CP, Wallis C, Melnick JL (1975) Fate of wastewater bacteria and viruses in soil. J Irrig Drain Div ASCE 101:157–174

Gilmour CM, Broadbent FE, Beck SM (1977) Recycling of carbon and nitrogen through land disposal of various wastes. In: Eliott LF, Stevenson FJ (eds) Soils for management of organic wastes and wastewaters. ASA, CSSA, SSSA, Madison, pp 171–194

Goh KM, Haynes RJ (1986) Nitrogen and agronomic practice. In: Haynes RJ (ed) Mineral nitrogen in the plant-soil system. Academic Press, New York, London, pp 379–468

Griffin RA, Jurinak JJ (1973) Estimation of activity coefficients from the electrical conductivity of natural aquatic systems and soil extracts. Soil Sci 116:26–30

Grigoropoulos SG, Smith JW (1971) Trace organics in subsurface waters. In: Faust SD, Hunter JV (eds) Organic compounds in aquatic environments. Dekker, New York, pp 95–118

Halevy J (1979) Fertilizer requirement for high cotton yields. Soils in Mediterranean type climates and their yield potential. Potash Inst. Bern, pp 359–365

Halevy J, Askozido A, Zarko S, Sagiv B, Natav Y (1975) Effect of nitrogen and phosphorus fertilizers on spring potatoes grown in the northern Negev. Hassadeh 58:916–920 (in Hebrew with English summary)

Halevy J, Feigin A, Marcovics T (1987) Interactive effects of N and P fertilizers on wheat. Hassadeh 66:2220–2222 (in Hebrew)

Hall GF, Wilding LP, Erickson EA (1976) Utilization and disposal of municipal industrial, and agricultural processing wastes on land. North Central Reg Committ NC-118, Wooster, Oh 2:1–8
Hanks RJ, Keller J, Rasmussen VP, Wilson GD (1976) Line source sprinkler for continuous variable irrigation – crop production studies. Soil Sci Soc Am Proc 40:426–429
Hanks RJ, James DW and Watts DW (1983) Irrigation management and crop production as related to nitrate mobility. In: Nelson DW, Elrick EE, Tanji KK (eds) Chemical mobility and reactivity in soil systems. Soil Sci Soc Am Madison Spec Pub 11:141–151
Harlin CC (1979) Land treatment methods in perspective. Prog Water Technol 11:1–14
Harmsen K, Bolt GH (1982) Movement of ions in soil. 1: Ion exchange and precipitation, 2: Ion exchange and dissolution. Georderma 28:85–116
Hauck RD (1984a) Nitrogen fertilizer effects on nitrogen cycle processes. In: Clark FE, Rosswall T (eds) Terrestrial nitrogen cycles. Swed Nat Sci Res Counc (Stockholm) Ecol Bull 33:551–562
Hauck RD (ed) (1984b) Nitrogen in crop production. Am Soc Agron, Madison 804 pp
Haynes RJ (1986) Mineral nitrogen in the plant-soil system. Academic Press, New York
Hill DE, Sawhney BL (1981) Removal of phosphorus from wastewater by soil under aerobic and anaerobic conditions. J Environ Qual 10:401–405
Hoff JD, Nelson DW, Sutton AL (1981) Ammonia volatilization from liquid swine manure applied to cropland. J Environ Qual 10:90–95
Hoffman G, van Genuchten MT (1983) Soil properties and efficient water use: water management for salinity control. In: Taylor HM, Jordan W, Sinclair T (eds) Limitations to efficient water use in crop production. Am Soc Agron, Madison, pp 73–85
Hook JE (1983) Movement of phosphates and nitrogen in soil following application of municipal wastewater. In: Nelson DW, Elrick DE, Tanji KK (eds) Chemical mobility and reactivity in soil systems. Soil Sci Soc Am Madison Spec Publ 11:241–255
Huang JC (1982) Implementing direct water reuse. In: Middlebrooks EJ (ed) Water reuse. Science Publisher, Ann Arbor, pp 265–276
Hunter JV (1971) Origin of organics from artificial contamination. In: Faust SD, Hunter JV (eds) Organic compounds in aquatic environments. Dekker New York, pp 51–94
Hunter JV, Kotalik TA (1974) Chemical and biological quality of treated sewage effluents. In: Sopper WE, Kardos LT (eds) Conference on recycling treated municipal wastewater through forest and cropland. US Environmental Protection Agency EPA-660/2-74-003, pp 6–27
Idelovitch E (1978) Wastewater reuse by biological-chemical treatment and groundwater recharge. J Water Pollut Control Fed 50:2723–2740
Idelovitch E, Michail M (1981) Gross organics measurements for monitoring of wastewater treatment and reuse. In: WJ Cooper (ed) Chemistry of water reuse, vol 1. Science Publisher, Ann Arbor, pp 35–64
Idelovitch E, Roth T, Michail M, Cohen A, Friedman R (1976) Advanced treatment, groundwater discharge and reuse of municipal wastewater. Dan Region Pilot Plant. TAHAL – Water Planning for Israel, Tel Aviv
Ingvalson RD, Rhoades JD, Page AL (1976) Correlation of alfalfa yield with various index of salinity. Soil Sci 122:145–153
Iskander IK (ed) (1981) Modelling wastewater renovation, land treatment. John Wiley & Sons, New York 802 pp
Iskandar IK, Syers JK (1980) Effectiveness of land application for phosphorus removal from municipal wastewater at Manteca, California, J Environ Qual 9:616–621
Israel Ministry of Health (ed) (1979) Recommendations for treatment of wastewater to be used for irrigation, 2nd draft
Jackson ML (1964) Chemical composition of soils. In: Bear FE (ed) Chemistry of the soil, 2nd edn. Van Nostrand Reinhold, New York, pp 71–141
Jansson SL, Persson J (1982) Mineralization and immobilization of soil nitrogen. In: Stevenson FJ (ed) Nitrogen in agricultural soils. Am Soc Agron, Madison. Agronomy 22:229–252
Jenkins TF, Martell CJ (1979) Pilot scale study of overland flow land treatment in cold climates. Prog Water Technol 11:207–214
Jensen ME (ed) (1974) Consumptive use of water and irrigation water requirements. Am Soc Civ Eng, New York, 215 pp

Jensen ME (1980) Design and operation of farm irrigation systems. Am Soc Agric Eng, POB 410, St Joseph, Mich 49085, USA

Jensen ME, Haise HR (1963) Estimating evapotranspiration from solar radiation. J Irrig Drain Div ASCE 89:15–41

Jones CA, Cole CV, Sharpley AN, Williams JR (1984) A simplified soil and plant phosphorus model, 1: Documentation. Soil Sci Soc Am J 48:800–805

Jones LHP, Jarvis SC (1981) The fate of heavy metals. In: Greenland DJ, Hayes MHB (eds) The chemistry of soil processes. John Wiley & Sons, New York Chichester, pp 593–620

Jurinak JJ, Santillan-Medrano J (1974) The chemistry and transport of lead and cadmium in soils. Utah Agric Exp Stn, Logan, Res Rep 18

Jury WA (1982) Simulation of solute transport with a transfer function model. Water Resour Res 18:363–368

Jury WA (1985) Spatial variability of soil physical parameters in solute migration: a critical literature review. Top Rep EA4 228, Elect Power Res Inst, Palo Alto Calif, 80 pp

Jury WA, Sposito G (1985) Field calibration and validation of solute transport models for the unsaturated zone. Soil Sci Soc Am J 49:1331–1341

Jury WA, Spencer WF, Farmer WJ (1983) Behavior assessment model for trace organics in soil. 1: Model description. J Environ Qual 12:558–564

Jury WA, Farmer WJ, Spencer WF (1984a) Behavior assessment model for trace organics in soil. 2: Chemical classification and parameter sensitivity. J Environ Qual 13:567–572

Jury WA, Spencer WF, Farmer WJ (1984b) Behavior assessment model for trace organics in soil. 3: Application of screening model. J Environ Qual 13:573–579

Jury WA, Spencer WF, Farmer WJ (1984c) Behavior assessment model for trace organics in soil. 4: Review of experimental evidence. J Environ Qual 13:580–586

Jury WA, Sposito G, White RE (1986) A transfer function model of solute transport through soil. I. Fundamental concepts. Water Resour Res 22:243–247

Kaddous FGA, Stubbs K, Morgans A (1986) Recycling of secondary treated effluent through vegetables and a loamy sand soil. Dept Agric Rural Aff, Victoria, Aust, Res Rep Ser 3

Kamprath EJ, Watson ME (1980) Conventional soil and tissue tests for assessing the phosphorus status of soils. In: Khasawneh FE, Sample EC, Kamprath EJ (eds) The role of phosphorus in agriculture. Am Soc Agron, Madison, pp 433–469

Kardos LT (1967) Waste water renovation by the land – a living filter. In: Brady NC (ed) Agriculture for the quality of our environment. Am Assoc Adv Sci 85:241–250

Kardos LT, Hook JE (1976) Phosphorus balance in sewage effluent treated soils. J Environ Qual 5:87–90

Kardos LT, Sopper WE, Myers EA, Parizek RR, Nesbitt JB (1974) Renovation of secondary effluent for reuse as a water resource. Environ Protect Technol Ser, US EPA-600/2-74-016

Kardos LT, Scarsbrook CE, Volk VV (1977) Recycling elements in wastes through soil-plant systems. In: Eliott LE, Stevenson FJ (eds) Soils for management of organic wastes and waste waters. Am Soc Agron, Madison, pp 301–324

Keefer RF, Singh RN, Horvath DJ (1986) Chemical composition of vegetables grown on an agricultural soil amended with sewage sludges. J Environ Qual 15:146–152

Keeney DR (1982a) Nitrogen management for maximum efficiency and minimum pollution. In: Stevenson JF (ed) Nitrogen in agricultural soils. Am Soc Agron Madison, Agronomy 22:605–649

Keeney DR (1982b) Nitrogen-availability indices. In: Page AL, Miller RH, Keeney DR (eds) Methods of soil analysis, 2nd edn, pt 2: Chemical and microbiological properties. Am Soc of Agron Madison, Igonomy 9:711–733

Kemp MS, Filip DS, George DB (1979) Overland flow and slow rate systems to upgrade wastewater lagoon effluent. Prog Water Technol 11:227–256

Kemper WD, van Schaik JC (1966) Diffusion of salts in clay-water systems. Soil Sci Soc Am Proc 30:534–540

Keren R, Bingham FT (1985) Boron in water, soils, and plants. Advances in soil science. Springer, Berlin Heidelberg New York, pp 229–276

Keren R, Bingham FT, Rhoades JD (1985) Effect of clay content in soil on boron uptake and yield of wheat. Soil Sci Soc Am J 49:1466–1470

Kidder EH (1976) Selection of the system for wastewater application on agricultural land. In: Knezek BD and Miller RH (eds) Application of sludges and waste waters on agricultural land: a planning and educational guide. Ohio Agric Res Dev Cent, Wooster, Oh, Res Bull 1090:8.1–8.7

King LD (1982) Land application of untreated industrial wastewater. J Environ Qual 11:638–644

Kipnis T, Feigin A, Vaisman I, Shalhevet J (1981) Wastewater application, dry matter production and nitrogen balance of Rhodes grass grown in fine-textured soil or on sand dunes. In: Proc 14th Int Grassland Congr, Kentucky, pp 288–291

Klute A (ed.) (1986) Methods of soil analysis, pt 1. Am Soc Agron, Madison Agronomy 9:687–734

Koerner EL, Haws DA (1979) Long-term effects of land application of domestic wastewater. Roswell, New Mexico. US EPA-600/2-79-047. Available from NTIS. Nat Tech Inf Cent, Springfield, Virg 22161

Korte NE, Skopp J, Fuller WH, Niebla EE, Alesii BA (1976) Trace element movement in soils: influence of soil physical and chemical properties. Soil Sci 122:350–359

Kott Y (1970) Chlorination of sewage oxidation pond effluents. In: Shuval H (ed) Developments in water quality research. Science Publisher, Ann Arbor, pp 189–197

Kowal NE (1983) An overview of public health effects. In: Page AL, Gleason TL, Smith JE, Iskanolar IK, Sommers LE (eds) Utilization of municipal wastewater and sludge on land. Univ Cal Press, Riverside, pp 329–395

Kristensen KK, Bond GJ (1977) The current status of bacterial and other pathogenic organisms in municipal wastewater and their potential health hazards with regard to agricultural irrigation. In: Ditri FM (ed) Wastewater renovation and reuse. Dekker, New York, pp 387–419

Lagerwerff JV, Holland JP (1960) Growth and mineral content of carrots and beans as related to varying osmotic and ionic composition effects in saline-sodic sand culture. Agron J 52:603–608

Lance JC (1972) Nitrogen removal by soil mechanisms. J Water Pollut Control Fed 44:1352–1361

Lance JC (1977) Phosphate removal from sewage water by soil columns. J Environ Qual 6:279–284

Langelier WF (1936) The analytical control of anticorrosion water treatment. J Am Water Works Assoc 28:1500–1521

Lanyon LE, Smith FW (1985) Potassium nutrition of alfalfa and other forage legumes: temperate and tropical. In: Munson RD (ed) Potassium in agriculture. Am Soc Agron Madison, pp 861–893

Larkin EP, Tierney JT, Sullivan R (1976) Persistance of viruses on sewage irrigated vegetables. J Environ Eng Div ASCE EEI 102:29–35

Latterell JJ, Dowdy RH, Clapp CE, Larson WE, Linden DR (1982) Distribution of phosphorus in soils irrigated with municipal wastewater effluent: 5 year study. J Environ Qual 11:124–128

Lau SL, Ekern PC, Loh PCS, Young RHF, Burbank NC, Dugan GL (1975) Water recycling of sewage effluents by irrigation: a field study in Ohau. Water Res Cent, Univ Hawaii, Honolulu, Tech Rep 79:110 pp

Letey J, Blair JM, Devitt D, Lund LJ, Nash P (1977) Nitrate-nitrogen in effluent from agricultural tile drains in California. Hilgardia 45:289–319

Linden DR, Clapp CE, Dowdy RH (1983) Hydrologic and nutrient management aspects of municipal wastewater and sludge utilization on land. In: Page AL, Gleason TL, Smith JE Jr, Iskandar IK, Sommers LE (eds) Utilization of municipal wastewater and sludge on land. Univ Cal Press, Riverside, pp 79–103

Lindsay WL (1973) Inorganic reactions of sewage wastes with soil. In: recycling municipal sludges and effluents on land. In: Proc Joint Conf Natl Assoc State Univ Land – Grant Colleges, Washington, DC, pp 91–96

Lindsay WL (1979) Chemical equilibria in soils. John Wiley & Sons, New York

Loehr RC (ed) (1977) Land as a waste management alternative. Science Publisher, Ann Arbor, 811 pp

Logan TJ, Chaney R (1983) Metals. In: Page AL, Gleason TL, Smith JE Jr, Iskandar IK, Sommers LE (eds) Utilization of municipal wastewater and sludge on land. Univ Cal Press, Riverside, pp 235–326

Luthin JN (1965) Drainage engineering. John Wiley & Sons, New York, 250 pp

Maas EV (1986) Salt tolerance of plants. Appl Agric Res 1:12–26

Maas EV, Hoffman GJ (1977) Crop salt tolerance—current assessment. J Irrig Drain Div ASCE 103:115–134

Maas EV, Clark RA, Francois LE (1982) Sprinkling induced foliar injury to pepper plants: effects of irrigation frequency, duration and water composition. Irrig Sci 3:101–109

Manka J, Rebhun M, Mandelbaum A, Bortinger A (1974) Characterization of organics in secondary effluents. Environ Sci Technol 8:1017–1020

Martin JP, Focht DD (1977) Biological properties of soils. In: Elliot LF, Stevenson FJ (eds) Soils for management of organic wastes and wastewaters. ASA CSSA SSSA, Madison, pp 115–119

Maynard DN, Barker AV, Minoti PL, Peck NH (1976) Nitrate accumulation in vegetables. Adv Agron 28:71–118

McBride MB (1989) Reactions controlling heavy metal solubility in soils. Adv Soil Sci 10:1–57

McNeal BL, Coleman NT (1966) Effect of solution composition on soil hydraulic conductivity. Soil Sci Soc Am Proc 30:308–312

McNeal BL, Pratt PF (1978) Leaching of nitrate from soils. In: Pratt PF (ed) Management of nitrogen in irrigated agriculture. Dep Soil Environ Sci, Univ Cal, Riverside, pp 195–230

McPherson JB (1979) Land treatment of wastewater at Werribee: past, present and future. Prog Water Technol 11: 15–31

Meiri A (1984) Plant response to salinity: application of experimental results. In: Shainberg I, Shalhevet J (eds) Soil salinity under irrigation – processes and management. Ecological studies. Springer, Berlin, Heidelberg New York, pp 284–297

Meisinger JJ (1984) Evaluating plant – available nitrogen in soil-crop systems. In: Hauck RD (ed) Nitrogen in crop production. Am Soc Agron, Madison, pp 391–416

Mengel K (1985) Dynamics and availability of major nutrients in soils. Adv Soil Sci 2:67–131

Mengel K, Kirkby EA (1987) Principles of plant nutrition. 4th edn. Int Potash Inst, Worblaufen-Bern, Switz

Metcalf & Eddy Inc (ed) (1979) Wastewater engineering: treatment, disposal reuse. McGrawhill, New York

Metzger L, Yaron B, Mingelgrin U (1983) Soil hydraulic conductivity as affected by physical and chemical properties of effluent. Agronomie 3:771–779

Miller RH (1974) The soil as a biological filter. In: Sopper WE, Kardos LT (eds) Conf Recycling treated municipal wastewater through forest and cropland. US Environmental Protection Agency EPA-660/2-74-003, pp 73–94

Montgomery JM (1985) Water treatment principles and design. Wiley Interscience, New York

Mortvedt JJ, Giordano PM, Lindsay WL (eds) (1972) Micronutrients in agriculture Proc, Conf Muscle Shoals, Alabama April 20–22, 1971, Soil Sci Soc Am, Madison, pp666

Munson RD (ed) (1985) Potassium in agriculture. Am Soc Agron, Madison Proc Int Symp Atlanta Georgia 7–10 July, 1223 pp

Nagpal NK (1985) Long-term phosphorus sorption in a brunisol in response to dosed effluent loading. J Environ Qual 14:280–285

Nagpal NK (1986) Effect of soil and effluent characteristics on phosphorus sorption in dosed columns. J Environ Qual 15:73–78

Nakayama FS (1982) Water analysis and treatment techniques to control emitter plugging. In: Proc Irrigation association exposition Conf, Portland, pp 97–112

Nakayama FS, Bucks DA (1981) Emitter clogging effects on trickle irrigation uniformity. ASAE Trans 24(1):77–80

Nakayama FS, Bucks DA (1985) Temperature effects on calcium carbonate precipitation. In: Proc 3rd Int Drip/trickle irrigation Cong. ASAE, Fresno California, vol 1: 45–49

Nakayama FS, Bucks DA (eds) (1986) Trickle irrigation for crop production. Elsevier, Amsterdam, 383 pp

Neal RH, Sposito G (1986) Effects of soluble organic matter and sewage sludge amendments on cadmium sorption by soils at low cadmium concentrations. Soil Sci 142:164–172

Nelson DW (1982) Gaseous losses of nitrogen other than through denitrification. In: Stevenson FJ (ed) Nitrogamin agricultural soils. Am Soc Agron, Madison. Agronomy 22:327–363

Nielsen DR, Biggar JW, Wierenga PJ (1982) Nitrogen transport processes in soil. In: Stevenson FJ (ed) Nitrogen in agricultural soils. Am Soc Agron, Madison. Agronomy 22:423–448

Nielsen DR, van Genuchten MT, Biggar JW (1986) Water flow and solute transport processes in the unsaturated zone. Water Resour Res 22:895–1085

Nommik H, Vahtras K (1982) Retention and fixation of ammonium and ammonia in soils. In: Stevenson FJ (ed) Nitrogen in agricultural soils. Am Soc Agron, Madison. Agronomy 22:123–171

Norvell WA (1972) Equilibria of metal chelates in soil solution. In: Mortvedt JJ, Giordano PM, Lindsay WL (eds) Micronutrients in agriculture. Soil Sci Soc Am Madison, pp 115–138

Noy J, Feinmesser A (1977) The use of wastewater for agricultural irrigation. In: Shuval HI (ed) Water renovation and reuse. Academic Press, New York, London, pp 73–92

NRC (National Research Council – ed) (1980) Mineral tolerance of domestic animals. Natl Acad Sci, Washington, 577 pp

Nye PH, Tinker PB (1977) Solute movement in the soil-root system. Univ Cal Press, Berkeley, pp 342

Olsen SR, Barber SA (1977) Effect of waste application on soil phosphorus and potassium. In: Eliott LF, Stevenson FJ (eds) Soils for management of organic wastes and wastewaters. ASA CSSA SSSA, Madison, pp 197–215

Olsen SR, Sommers LE (1982) Phosphorus. In: Page AL, Miller RH, Keeney DR (eds) Methods of soil analysis, 2: Chemical and microbiological properties. Am Soc Agron, Madison. Agronomy 9: 403–430

Olson RA, Kurtz LT (1982) Crop nitrogen requirements, utilization, and fertilization. In: Stevenson FJ (ed) Nitrogen in agricultural soils. Am Soc Agron Madison, pp 567–604

Oron G, De Malach J (1987) Reuse of domestic wastewater for irrigation in arid zones: a case study. Water Resour Bull 23:777–783

Oron G, Ben-Asher J, DeMalach Y (1982) Effluent in trickle irrigation of cotton in arid zones. J Irrig Drain Div 108:115–126

Oron G, De Malach J, Hoffman Z (1988) Seven successive seasons of subsurface dripper irrigation using effluent on field crops. Water Irrig Rev 8:4–8

Oster JD, Hoffman GJ, Robinson FE (1984) Management alternatives: crop, water, soil. Cal Agric 38(10):29–32

Oster JD, Tracy JE, Meyer JL, Snyder MJ (1989) Selenium in or near the southern coastal range: well water and vegetables crops. In: Tanji KK, Valoppi L, Woodring RC (eds) Selenium contents in animal and human food crops grown in California. Univ Cal Publ 3330:51–55

Overcash MR (1983) Land treatment of municipal effluent and sludge: Specific organic compounds. In: Page AL, Gleason TL, Smith JE Jr, Iskandar IK, Sommers LE (eds) Utilization of municipal wastewater and sludge on land. Univ Cal Press, Riverside, pp 199–231

Overman AR (1975) Effluent irrigation of pearl millet. J Environ Eng Div ASCE 102:193–199

Overman AR, Nguy A (1975) Growth response and nutrient uptake by forage under effluent irrigation. Commun Soil Sci Plant Anal 9:81–93

Overman AR, Ku HC (1976) Effluent irrigation of rye and rye grass. J Environ Eng Div ASCE 102:475–483

Page AL (1974) Fate and effects of trace elements in sewage sludge when applied to agricultural lands. A literature review study. US Environ Protect Ag EPA-670-/2-74-005

Page AL, Chang AC (1985) Fate of wastewater constituents in soil and groundwater: trace elements. In: Pettygrove GS, Asano T (eds) Irrigation with reclaimed municipal wastewater – a guidance manual. Lewis, Chelsea, pp 13.1–13.16

Page AL, Chang AC, Sposito G and Mattigod S (1981) Trace elements in wastewater: their effects on plant growth and composition and their behavior in soils. In Iskander IK (ed) Modelling wastewater renovation, land treatment. John Wiley & Sons, New York, pp 182–222

Page AL, Miller RH, Keeney DR (eds) (1982) Methods of soil analysis, pt 2: Chemical and microbiological properties, 2nd edn. Am Soc Agron, Madison. Agronomy 9

Page AL, Gleason TL, Smith JE, Iskander IK, Sommers LE (eds) (1983) Utilization of municipal wastewater and sludge on land. Univ Cal Press, Riverside, 480 pp

Page AL, Logan TJ, Ryan JA (eds) (1987) Land application of sludge – food chain implications. Lewis, Chelsea 168 pp

Pahren HR, Jakubowski W (eds) (1980) Wastewater aerosols and disease. US Environ Protect Ag EPA-60/0-80-028

Palazzo AJ (1977) Land application of wastewater: forage growth and utilization of applied nitrogen, phosphorus and potassium. In: Loehr RC (ed) Land as a waste management alternative. Science Publisher, Ann Arbor, pp 171–180

Papadopaulos I, Stylianou Y (1988) Trickle irrigation of cotton with treated sewage effluent. J Environ Qual 574–580

Parfitt RL (1978) Anion adsorption by soils and soil materials. Adv Agron 30:1–50

Parizek RR, Kardos LT, Sopper WE, Meyers EA, Davis DE, Farell MA, Nesbit JB (1967) Wastewater renovation and conservation. Penn State Stud 23. Penn State Univ, University Park, 71 pp

Parr JF (1973) Chemical and biochemical considerations for maximizing the efficiency of fertilizer nitrogen. J Environ Qual 2:75–84

Parson D, Brownlee C, Wetler D, Maurer A, Haughton E, Kornder L, Selzak M (1975) Health aspects of sewage effluent irrigation. Pollut Control Branch, B C Water Resour Serv Dept Lands For Water Resour, Victoria, BC, Can

Payne WJ (1981) Denitrification. John Wiley & Sons, New York

Pearson GA (1960) Tolerance of crops to exchangeable sodium. USDA Inf Bull 216:4 pp

Penman HL (1963) Vegetation and hydrology. Tech Commun 53 Common Bur Soils, Harpenden, UK, 125 pp

Pescod MB (1986) Integration of urban wastewater treatment and effluent re-use in irrigation. Water Pollut Control Fed 85:459–471

Pescod MB, Alka U (1984) Urban effluent reuse for agriculture in arid and semi-arid zones. In: Reuse of sewage effluent. Telford, London, pp 93–106

Pettygrove GS, Asano T (eds) (1985) Irrigation with reclaimed municipal wastewater – a guidance manual. Lewis, Chelsea

Phene CJ, Howell TA, Sikorski MD (1985) A traveling trickle irrigation system. Adv Irrig 3:2–49

Pound CE, Crites RW (1973a) Characteristics of municipal effluents. In: Recycling municipal sludges and effluents on land. In: Proc Conf Natl Assoc State Univ; and Land-Grant Colleges, Washington, pp 49–61

Pound CE, Crites RW (1973b) Wastewater treatment and reuse by land application, vols 1,2. US Environ Protect Ag EPA-660/2-73-006a, b

Pound CE, Griffes DA, Crites RW (1983) Engineering and economics of applying wastewater and sludge to land. In: Page AL, Gleason TL, Smith JE, Iskander IK, Sommers LE (eds) Utilization of municipal wastewater and sludge on land. Univ Cal Press, Riverside, pp 53–75

Pratt PF, Lund LJ, Rible JM (1978) An approach to measuring leaching of nitrate from freely drained irrigated fields. In: Nielsen DR, McDonald JC (eds) Nitrogen in the environment, Academic Press, New York London, pp 223–256

Purvis D (1985) Trace element contamination in the environment, rev edn. Elsevier, Amsterdam

Quin BF, Syers JK (1978) Surface irrigation of pasture with treated sewage effluent, 3: Heavy metal content of sewage effluent, sludge, soil, and pasture, NZ J Agric Res 21:435–442

Quin BF, Woods PH (1978) Surface irrigation of pasture with treated sewage effluent, 1: nutrient status in soil and pasture NZ J Agric Res 21:419–426

Rao PSC, Jessup RE (1983) Sorption and movement of pesticides and other toxic organic substances in soil. In: Nelson DW, Elric EE, Tanji KK (eds) Chemical mobility and reactivity in soil systems. Soil Sci Soc Am Madison Spec Publ 11:183–201

Rappaport BD, Martens DC, Reneau RB, Simpson TW (1987) Metal accumulation in corn and barley grown on sludge amended Typic Ochraqualf. J Environ Qual 16:29–33

Rebhun M, Manka J (1971) Classification of organics in secondary effluents. Environ Sci Technol 5:606–609

Reed SC (1982) Health effect and land application of wastewater. In: Middlesbrook EJ (ed) Water reuse. Science Publishers, Ann Arbor, pp 753–782

Reisenauer HM (ed) (1978) Soil and plant-tissue testing in California, rev edn. Div Agric Sci Univ Cal Bull 1979

Rhoades JD (1974) Drainage for salinity control. In: van Schilfgaarde J (ed) Drainage for agriculture. Am Soc of Agron, Madison. Agronomy 17:433–461

Rhoades JD (1984) Principles and methods of monitoring soil salinity. In: Shainberg I, Shalhevet J (eds) Soil salinity under irrigation – processes and management. Springer Ecol Stud 51:130–142

Rhoades JD, Oster JD (1986) Solute control. In: Klute D (ed) Methods of soil analysis. I. Physical and mineralogical methods, 2nd edn. Am Soc Agron, Madison. Agronomy 9: 985–1005

Rible JM, Pratt PF, Lund LJ, Holtzclaw KH (1979) Nitrates in the unsaturated zone of freely drained fields. In: Pratt PF (ed) Nitrate in effluents from irrigated land. Final Rep US Nat Sci Found, Univ Cal, pp 297–320

Roberts S, McDole RE (1985) Potassium nutrition of potatoes. In: Munson RD (ed) Potassium in agriculture. Am Soc Agron Madison, pp 799–818

Robinson DL (1985) Potassium nutrition of forage grasses. In: Munson RD (ed) Potassium in agriculture. Am Soc Agron Madison. pp 895–914

Ryden JC, Pratt PF (1980) Phosphorus removal from wastewater applied to land. Hilgardia 48(1):1–36

Ryden JC, Syers JK, Iskander IK (1981) Evaluation of a simple model for predicting phosphorus removal by soils during land treatment of wastewater. In: Iskander IK (ed) Modeling wastewater renovation, land treatment, John Wiley & Sons, New York, pp 647–667

Sabol GV, Bouwer H, Wierenga PJ (1987) Irrigation effects in Arizona and New Mexico. J Irrig Drain Div ASCE 113:30–48

Sanchez Duron N (1985) Mexican experience in using sewage effluent for large-scale irrigation. In: FAO (ed) Reg Sem Treatment and use of sewage effluent for irrigation. FAO, Rome. 10 pp

Sauerbeck D (1987) Effects of agricultural practices on the physical, chemical and biological properties of soils, pt 2: Use of sewage sludge and agricultural wastes. In: Barth H, L'hermite P (eds) Scientific basis for soil protection in the European Community. Elsevier, New York, pp 181–210

Sawhney BL (1977) Predicting phosphate movement through soil columns. J Environ Qual 6:86–89

Sawhney BL, Hill DE (1975) Phosphate sorption characteristics of soils treated with domestic wastewaters. J Environ Qual 4:342–346

Schirado T, Vergara I, Schalscha EB, Pratt PF (1986). Evidence for movement of heavy metals in a soil irrigated with untreated wastewater. J Environ Qual 15:9–12

Schmidt EL (1982) Nitrification in soil. In: Stevenson FJ (ed) Nitrogen in agricultural soils. Am Soc Agron, Madison. Agronomy 22:253–288

Sepp E (1971) The use of sewage for irrigation – a literature review. Cal Dept Health, Bur Sanit Eng, Berkeley, Cal

Shainberg I, Letey J (1984) Response of soils to sodic and saline conditions – a review. Hilgardia 52:1–57

Shainberg I, Oster JD (1978) Quality of irrigation water, Int Irrig Inf Cent (IIIC), Bet Dagan, Isr, Publ 2:65 pp

Shalhevet J (1965) Drainage of fine textured soils-Hazorea experiment. Volcani Inst, ARO, Spec Rep, 94 pp (Hebrew with English summary)

Shalhevet J (1970) The use of saline water for irrigation. In: Gerson D (ed) Dialogue and development. 2nd World Congr Engineers and architects, Tel Aviv, pp 437–442

Shalhevet J (1973) Irrigation with saline water. In: Yaron B, Danfors E, Vaadia Y (eds) Arid zone irrigation. Springer, Berlin, Heidelberg New York, pp 263–276

Shalhevet J (1974) Aspects of soil salinity and sodicity in relation to irrigation and reclamation. Dep Sci Publ, Volcani Cent, ARO, Bet Dagan, Isr, Spec Publ 39

Shalhevet J, Bernstein L (1968) Effect of vertically heterogeneous soil salinity on plant growth and water uptake. Soil Sci 106:85–93

Shalhevet J, Bielorai H (1978) Crop water requirements in relation to climate and soil. Soil Sci 125:240–247

Shalhevet J, Kamburov J (1973) Irrigation and salinity, a worldwide survey. Int Comm Irrig Drain, New Delhi, 106 pp

Shalhevet J, Mantell A, Bielorai H, Shimshi D (1981) Irrigation of field and orchard crops under arid conditions, rev edn., Int Irrig Inf Center (IIIC), Bet Dagan, Isr, Publ 1: 132 pp

Sharpley AN, Jones CA, Gray C, Cole CV (1984) A simplified soil and plant phosphorus model, 2: Prediction of labile, organic, and sorbed phosphorus. Soil Sci Soc Am J 48:805–809

Shelef G (1977) Water reuse in Israel. In: Shuval HI (ed) Water renovation and reuse. Academic Press, New York London, pp 311–332

Shuval HI (1975) Disinfection of wastewater for agricultural utilization. Prog Water Technology 1: 857–867

Shuval HI (ed) (1977) Health considerations in water renovation and reuse. In: Water renovation and reuse. Academic Press, New York London, pp 33–72
Shuval HI (1978) Land treatment of wastewater in Israel. In: Proc Land Treatment symposium, USA CRREL, Hanover, NH, pp 429–436
Shuval HI (1987) Wastewater reuse for irrigation: evolution of health standards. WHO Water Qual Bull 12:79–84
Shuval HI, Adin A, Fattal B, Rawitz E, Yekutiel P (1986) Wastewater irrigation in developing countries. Health effects and technical solutions. World Bank Tech Pap 51, 324 pp
Smith JH, Peterson JR (1982) Recycling of nitrogen through land application of agricultural, food processing and municipal wastes. In: Stevenson FJ (ed) Nitrogen in agricultural soils. Am Soc Agron, Madison. Agronomy 22:791–831
Smith WH, Evans JO (1977) Special opportunities and problems in using forest soils for organic waste application. In: Eliott LF, Stevenson FJ (ed) Soils for management of organic wastes and wastewaters. ASA SSSA CSA, Madison, pp 429–454
Sommers LE, Nelson DW (1976) Analyses and their interpretation for sludge application to agricultural land. In: Knezek BD, Miller RH (eds) Application of sludge and wastewater to agricultural land: a planning and educational guide. Ohio Agric Res Dev Cent, Wooster, Res Bull 1090, pp 3.1–3.7
Sommers LE, Sutton AL (1980) Use of waste materials as sources of phosphorus. In: Khasawneh FE, Sample EC, Kamprath EJ (eds) The role of phosphorus in agriculture. Am Soc of Agron, Madison, pp 515–544
Sommers LE, Nelson DW, Owens LB (1979) Status of inorganic phosphorus in soils irrigated with municipal wastewater. Soil Sci 127:340–350
Sommers L, van Volk V, Giordano PM, Sopper WE, Bastian R (1987) Effects of soil properties on accumulation of trace elements by crops. In: Page AL, Logan TG, Ryan JA (eds) Land application of sludge. Lewis, Chelsea, pp 5–24
Sopper WE, Kardos LT (eds) (1974) Conference on recycling treated municipal wastewater through forest and cropland. US Environ Protect Ag EPA-660/2-74-003
Sparks DL, Huang PM (1985) Physical chemistry of soil potassium. In: Munson RD (ed) Potassium in agriculture. Am Soc Agron, Madison, pp 201–276
Stanford G (1973) Rationale for optimum nitrogen fertilization in corn production. J Environ Qual 2:159–166
Stanford G (1982) Assessment of soil nitrogen availability. In: Stevenson FJ (ed) Nitrogen in agricultural soils. Am Soc Agron Madison. Agronomy 22:651–688
Stevenson FJ (1982a) Humus chemistry, John Wiley & Sons, New York
Stevenson FJ (1982b) Organic forms of soil nitrogen. In: Stevenson FJ (ed) Nitrogen in agricultural soils. Am Soc Agron, Madison. Agronomy 22:67–122
Stevenson FJ (1985) Cycles of soil, carbon, nitrogen, phosphorus sulfur, micronutrients. John Wiley & Sons, New York
Stevenson FJ, Ardakani MS (1972) Organic matter reactions involving micronutrients in soils. In: Mordvedt JJ, Giordano PM, Lindsay WL (eds) Micronutrients in agriculture. Soil Sci Soc Am Inc, Madison, pp 79–114
Strom AG (1984) Reuse of sewage effluent in Australia. In: Reuse of sewage effluents. Telford, London, pp 35–44
Stuanes AO (1984) Phosphorus sorption of soils to be used in wastewater renovation. J Environ Qual 13:220–224
Suarez DL (1981) Relation between pH_c and sodium adsorption ratio (SAR) and an alternative method of estimating SAR of soil or drainage waters. Soil Sci Soc Am J 45:469–475
Sullivan RH, Cohn MM, Baxter SS (1973) Survey of facilities using land application of wastewater. US Environ Protect Ag EPA-430/9-73-006
Syers JK, Iskander IK (1981) Soil-phosphorus chemistry. In: Iskander IK (ed) Modeling wastewater renovation, land treatment. John Wiley & Sons, New York, pp 571–599
Talsma T (1963) The control of saline groundwater. Meded Landbouwh Wageningen 63(10):1–68
Tanji KK (1982) Modeling of the soil nitrogen cycle. In: Stevenson JF (ed) Nitrogen in agricultural soils. Am Soc Agron, Madison. Agronomy 22:721–772

Taras MJ, Greenberg AE, Hoak RD, Rand MC (1971) Standard methods for the examination of water and wastewater, 13th edn. Am Public Health Assoc, Washington

Taylor AW, Kilmer VJ (1980) Agricultural phosphorus in the environment. In: Khasawneh FE, Sample EC, Kamprath EJ (eds) The role of phosphorus in agriculture. Am Soc Agron, Madison, pp 545–557

Thomas R, Law JP (1977) Properties of wastewaters. In: Eliott LF, Stevenson FJ (eds) Soils for management of organic wastes and wastewaters. SSSA ASA CSSA, Madison, pp 47–72

Tiller KG (1989) Heavy metals in soils and their environmental significance. Adv Soil Sci 9: 113–142

Tucker TC, Hauck RD (1978) Removal of nitrogen by various irrigated crops. In: Pratt PF (ed) Management of nitrogen in irrigated agriculture. Dep Soil Environ Sci, Univ Cal, Riverside, pp 135–167

Underwood EJ (1979) Environmental sources of heavy metals and their toxicity to man and animals. Prog Water Technology 11:33–45

USBR (US Bureau of Reclamation, US Department of Interior – ed) (1974) Design of small dams. A water resource technical publication. US Gov Print Off, Washington

US Salinity Laboratory Staff (ed) (1954) Diagnosis and improvement of saline and alkali soils. US Dept Agric Handbook 60:160 pp

USDA, SCS (eds) (1967) Irrigation water requirements. Tech Release 21, Washington

Vaisman I, Shalhevet J, Kipnis T, Feigin A (1981) Reducing ground-water pollution from municipal wastewater irrigation of Rhodes grass grown on sand dunes. J Environ Qual 10:434–439

Vaisman I, Shalhevet J, Kipnis T, Feigin A (1982) Water regime and N fertilization for Rhodes grass irrigated with municipal wastewater on sand dune soil. J Environ Qual 11:230–232

Vaisman I, Kipnis T, Sharabani N, Bielorai H, Feigin A (1986) Response of corn grown in sand dunes to irrigation with municipal wastewater effluents. Hassadeh 66:2498–2501 (Hebrew with English summary)

van Genuchten MT, Alves WJ (1982) A compendium of available analytical solutions of the one-dimensional convective-dispersive solute transport equation. USDA, SEA, Washington, Tech Bull 1661

van Schilfgaarde J (ed) (1974) Drainage for agriculture. Am Soc Agron, Madison. Agronomy 17, 700 pp

van Schilfgaarde J (1984) Drainage design for salinity control. In: Shainberg I, Shalhevet J (eds) Soil salinity under irrigation-processes and management. Springer, Berlin Heidelberg New York, pp 190–197

van Schilfgaarde J, Bernstein L, Rhoades JD, Rawlins SL (1974) Irrigation management for salt control. J Irrig Drain Div ASCE 100:321–338

Vlamis J, Williams DE, Corey JE, Page AL, Gange TJ (1985) Zinc and cadmium uptake by barley in field plots fertilized seven years with urban and suburban sludge. Soil Sci 139:81–87

Wagenet RJ (1983) Principles of salt movement in soils. In: Nielsen DW, Elrick EE, Tanji KK (eds) Chemical mobility and reactivity in soil systems. Soil Sci Soc Am Spec Publ 11:123–140

Walsh LM, Beaton JE (eds) (1973) Soil testing and plant analysis, rev edn. Soil Sci Soc Am, Madison

Warrick AW, Mullen GJ, Nielsen DR (1977) Scaling field-measured soil hydraulic properties using a similar media concept. Water Resour Res 13:355–362

Webber LR, Beauchamp EG (1977) Land utilization and disposal of organic wastes in cool subhumid and humid regions. In: Eliott LF, Stevenson FJ (eds) Soils for management of organic wastes and wastewaters. Am Soc Agron, Madison, pp 457–470

Welch LF, Flannery RL (1985) Potassium nutrition of corn. In: Munson RD (ed) Potassium in agriculture. Am Soc Agron Madison, pp 647–664

Wesseling J (1974) Crop growth and wet soils. In: van Schilfgaarde J (ed) Drainage for agriculture. Am Soc Agron, Madison. Agronomy 17:7–37

WHO (World Health Organization – ed) (1973) Reuse of effluents. Methods of wastewater treatment and health safeguards: report of a WHO Meeting of Experts. WHO Geneva Tech Rep Ser 517

WHO Scientific Group (ed) (1989) Health guidelines for the use of wastewater in agriculture and aquaculture: report of a WHO Scientific group. WHO Geneva Tech Rep Ser 778:74 pp

Williams DE, Vlamis J, Pukite AH, Corey, JE (1980) Trace element accumulation, movement and distribution in the soil profile from massive applications of sewage sludge. Soil Sci 129:119–132

Williams DE, Vlamis J, Pukite AH, Corey, JE (1984) Metal movement in sludge-treated soils following six years of sludge additions. 1. Cd, Cu, Pb, and Zn. Soil Sci 137:351–359

Williams DE, Vlamis J, Pukite AH, Corey JE (1985) Metal movement in sludge treated soils following six years of sludge additions: 2. Ni, Co, Fe, Mn, Cr and Hg. Soil Sci 140:120–125

Williams DE, Vlamis J, Pukite AH, Corey, JE (1987) Metal movement in sludge-amended soils: a nine-year study. Soil Sci 143:124–131

Witty JE, Flach KW (1977) Site selection as related to utilization and disposal of organic wastes. In: Eliott LF, Stevenson EJ (eds) Soils for management for organic wastes and wastewaters. Am Soc Agron, Madison, pp 327–345

Wolcott AR, Cook RL (1976) Crop and system management for wastewater application to agricultural land. In: Knezek BD, Miller RH (eds) Application of sludges and wastewaters on agricultural land: a planning and educational guide. Ohio Agric Res Dev Cent, Wooster, Res Bull 1090, pp 7.1–7.12

Woldendorp JW (1968) Losses of soil nitrogen. Stickstoff 12:32–46

Yaron B, Shalehvet J, Shimshi D (1973) Pattern of salt distribution in trickle irrigation. In: Hadas A (ed) Physical aspects of soil, water and salt in ecosystems. Springer Ecol Stud 4:389–394

Young EE (1980) Cost of maintaining public health standards for spray irrigation of municipal wastewater systems. J Environ Qual 9:354–358

Subject Index

MIX
Papier aus verantwortungsvollen Quellen
Paper from responsible sources
FSC® C105338

If you have any concerns about our products,
you can contact us on
ProductSafety@springernature.com

In case Publisher is established outside the EU,
the EU authorized representative is:
Springer Nature Customer Service Center GmbH
Europaplatz 3, 69115 Heidelberg, Germany

Printed by Libri Plureos GmbH
in Hamburg, Germany